Energy Myths and Realities

Energy Myths and Realities:

Bringing Science to the Energy Policy Debate

Vaclav Smil

The AEI Press

Publisher for the American Enterprise Institute

WASHINGTON, D.C.

Distributed by arrangement with the Rowman & Littlefield Publishing Group, 4501 Forbes Boulevard, Suite 200, Lanham, Maryland 20706. To order call toll free 1-800-462-6420 or 1-717-794-3800. For all other inquiries please contact AEI Press, 1150 Seventeenth Street, N.W. Washington, D.C. 20036 or call 1-800-862-5801.

Library of Congress Cataloging-in-Publication Data

Smil, Vaclav.
  Energy myths and realities : bringing science to the energy policy debate / Vaclav Smil.
      p. cm.
  Includes bibliographical references and index.
  ISBN-13: 978-0-8447-4328-8
  ISBN-10: 0-8447-4328-3
  1. Renewable energy sources. 2. Energy policy. I. Title.
  TJ808.S639 2010
  333.79'4—dc22

                                        2010009437
14 13 12 11 10   1 2 3 4 5 6 7

Printed in the United States of America

Homines libenter quod volunt credunt
Men believe what they want to

— Publius Terentius

Contents

List of Figures

Key to Units of Measure

°C	degree Celsius (unit of temperature)
cm	centimeter (unit of length)
cm^3	cubic centimeter
E	exa (prefix signifying 10^{18})
EJ	exajoule (quintillion joules; unit of energy)
G	giga (prefix signifying 10^9)
g/Wh	grams per watt hour (measure of energy density)
Gb	billion barrels
GJ	gigajoule (billion joules; unit of energy)
GL	billion liters
GW	gigawatt (billion watts; unit of power)
GW_e	gigawatt electric (billion watts of electric power)
h	hour
J	joule (unit of energy)
k	kilo (prefix signifying 10^3)
kg	kilogram (thousand grams; unit of mass)
km	kilometer (thousand meters; unit of length)
km/h	kilometers per hour (unit of speed)
km^3	cubic kilometers (unit of volume)
kW	kilowatt (thousand watts; unit of power)
kW_e	kilowatt electric (thousand watts of electric power)
K	degree Kelvin (unit of temperature)
L	liter (unit of volume)
L/km	liters per kilometer (unit of fuel consumption)
m	meter (unit of length)
m/s	meters per second (unit of speed)
m^2	square meter (unit of area)

m^3	cubic meter (unit of volume)
mpg	miles per gallon (unit of fuel consumption)
M	mega (prefix signifying 10^6)
MJ	megajoule (million joules; unit of energy)
Mbpd	million barrels per day
MW	megawatt (unit of power)
MWh	megawatt hour (unit of energy)
P	peta (prefix signifying 10^{15})
PWh	petawatt hour (quadrillion watt hours; unit of energy)
ppm	parts per million
T	tera (prefix signifying 10^{12})
Tb	trillion barrels
TW	terawatt (trillion watts; unit of power)
TW_e	terawatt electric (trillion watts of electric power)
TWh	terawatt hour (unit of energy)
W	watt (unit of power)
W/m^2	watts per square meter (unit of power density)
Wh/kg	watt hours per kilogram (unit of energy density)
Wh/km	watt hours per kilometer (measure of energy consumption)

Introduction

Modern civilization is the product of incessant large-scale combustion of coals, oils, and natural gases and of the steadily expanding generation of electricity from fossil fuels, as well as from the kinetic energy of water and the fissioning of uranium nuclei.[1] Yet, for many decades, this fundamental link between the rising use of energies and the growing complexity and greater affluence of human societies was overlooked both by the public and by policymakers. The public was not concerned about energy supplies; media coverage of energy matters was sporadic; and no major Western government had a ministry devoted specifically to energy affairs.

This lack of interest changed with what came to be known as the first energy crisis—the increase in oil prices driven by the Organization of the Petroleum Exporting Countries (OPEC) in 1973 and 1974. This rise, from less than $2/barrel in early 1973 to more than $11/barrel by the spring of 1974 (BP 2009), was deliberately engineered by the leading oil exporters and did not take place in response to any physical shortage of the fuel. It went further than originally intended, cutting short the unprecedented period of economic expansion following World War II. It also turned the attention of individuals, organizations, and governments to the increasingly challenging task of securing a sufficient supply of sensibly priced energy. Moreover, this challenge coincided with the genesis of a new environmental consciousness and, hence, with efforts to reduce environmental pollution and prevent further ecosystemic degradation.

Suddenly, everybody seemed to become an energy expert, eager to proffer solutions. In reality, however, only a relatively small group of people understood energy affairs well enough to recognize how much was unknown about the structure and dynamics of complex energy systems, and how perilous it was to prescribe any lasting course of action. Those

1

knowledge gaps were largely filled during the years of intensifying energy studies that followed the first and then the second round of oil price increases (1979–81). But after those subsequent prices collapsed—from the peak of almost $40/barrel in March 1981 to $20/barrel by January 1986, and to less than $10/barrel in April 1986—the complacency of the period before 1973 rapidly returned (BP 2009). Instant experts reoriented themselves toward other concerns, such as global warming, globalization, and the new microprocessor-based economy.

Lost Opportunities

Unfortunately, some sensible policies aimed at reducing wasteful energy use were completely (and indefensibly) abandoned at this time. The best American example of this irrational retreat was the fate of the Corporate Average Fuel Economy (CAFE) regulations. Incredibly, the typical efficiency of America's cars in the early 1970s was about the same as it had been in the early 1930s. Technical advances had brought huge efficiency gains to virtually every mode of common energy conversion, thanks to the introduction of transistors and integrated circuits, the adoption of fluorescent lights, improvements in massive two-stroke diesel engines in ships, the commercialization of jet engines and stationary gas turbines, and innovations in oil refining and in plastics and fertilizer production. But American-made cars of the early 1970s could still get only about 13 miles per gallon of gasoline, wasted at least 85 percent of the purchased fuel, and performed no better than they had before World War II—that is, deeply below the technical potential of the day (Sivak and Tsimhoni 2009).

New rules that came into effect in 1975 commendably doubled average efficiency from 13.5 to 27.5 mpg by 1985, but no further improvements followed until new legislation was adopted in 2008 (see figure I-1). This failure to pursue greater fuel efficiency was an irrational choice and, hence, an irresponsible policy. It came about because of low oil prices,[2] and it led to a higher dependence on imports: By 1990, America imported 47 percent of its crude oil, compared to 37 percent in 1973. At issue here is not domestic energy self-sufficiency,[3] but the enormous trade deficits created by oil imports that weaken the nation's currency and long-term security and affect

FIGURE I-1

TRENDS IN CORPORATE AVERAGE FUEL ECONOMY (CAFE)
FOR U.S. VEHICLES

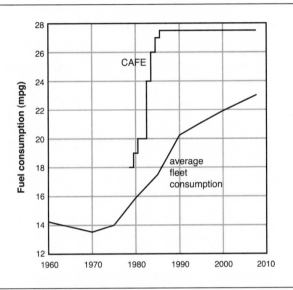

SOURCES: CAFE rates plotted from U.S. Congressional Research Service, Resources, Science, and Industry Division (2003); average performance calculated from U.S. Department of Transportation, Research and Innovative Technology Administration, Bureau of Transportation Statistics (2007, table 4-11, http://www.bts.gov/publications/national_transportation_statistics/html/table_04_11.html).

its strategic posture. In 2008, the United States bought 65 percent of its crude oil abroad, and the cost of imported oil and refined oil products was the single largest contributor—48 percent—to the country's more than $700 billion trade deficit.[4]

This lapse has been made much worse by the introduction of unnecessarily massive and mostly very inefficient vehicles—sport utility vehicles (SUVs), vans, and pickup trucks—that have been used overwhelmingly as passenger cars. Widespread ownership of two-axle, four-tire vehicles other than passenger cars[5] depressed the aggregate U.S. vehicle fleet performance to only about 22 mpg by 2006 (Sivak and Tsimhoni 2009). This low average mileage made little economic difference at the time, because energy prices remained low and fairly stable throughout the 1990s and for the first few years of the new century. During that period, energy supplies once again ceased to be a matter of major concern. Indeed, by 1998 the average price of crude oil had fallen to less than $12/barrel (a mere $16/barrel in 2008 dollars), and the oil industry's stocks were one of the worst performing investments of the 1990s.

Not until the early years of the new millennium, when oil prices began to rise once again, did attention return to energy supplies. During the latter half of 2003, the price of crude oil reached $25–$30/barrel, and during 2004 it came close to, and briefly even rose above, $40/barrel. The upward trend continued in 2005 and for the first eight months of 2006, and the media came to comment routinely on record high prices. In reality, no records were broken once two key price corrections—adjusting for the intervening inflation and taking into account lower oil intensity of Western economies[6]—were made. Until the early summer of 2008, these doubly adjusted oil prices remained well below the records set during the early 1980s.

In August 2006, the weighted mean price of all traded oil peaked at more than $71/barrel; it then fell by 15 percent within a month and closed the year at about $56/barrel. But during 2007, it again rose steadily. By November it reached almost $100/barrel in trading on the New York Mercantile Exchange (NYMEX; see figure I-2a), and during the first half of 2008 that price rose by half, reaching a high of $147.27/barrel on July 11. As always, prices for the basket of OPEC oils, including mostly heavier and more sulfurous crudes, remained lower (see figure I-2b).[7] But just three weeks after setting a record, oil prices fell by more than 20 percent, to about $115/barrel. By November 12 the price had fallen below $50/barrel, and a year later it was around $75, a rise largely caused by the falling value of the U.S. dollar. As always, any long-range forecast remains a guess, but, barring a prolonged global depression, nobody expects that oil prices will retreat to pre-2004 levels, because the latest round of energy concerns is driven by three powerful factors that will not disappear in the foreseeable future: widespread fears about an imminent peak of global crude oil extraction; apprehension about greater than usual political instability in the Middle East, largely a result of the Iran factor; and concerns about the socioeconomic repercussions of the quest to reduce greenhouse gas emissions (caps on carbon emissions and carbon taxes, for example).

No wonder that uninformed, and outright misinformed, pontification on oil futures has been reaching new heights, nor that this flood of opportunistic

FIGURE 1-2a

MONTHLY MEAN PRICES OF LIGHT SWEET CRUDE OIL ON NYMEX, 1999–2008

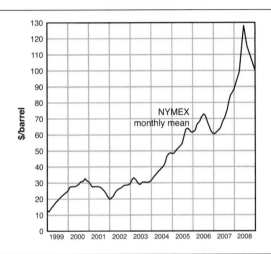

SOURCE: Plotted from Omega Research (1997).

FIGURE 1-2b

ANNUAL MEAN PRICES OF THE OPEC REFERENCE BASKET (ORB)

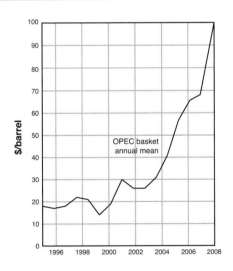

SOURCE: Plotted from OPEC data series, Organization of the Petroleum Exporting Countries (2009).

"analyses" and sensationalized "revelations" has been magnified by scores of cable TV news channels eager to fill their round-the-clock coverage with any willing talking head, and by the self-appointed experts of the blogosphere. Sources, claims, and opinions on energy matters are thus beyond counting, and, as might be expected, the rising quantity of the discourse has been inversely proportional to the quality of the disseminated information. As a result, the general public and policymakers alike have been assaulted by a wave of biased, misinterpreted, or outright false information.

Ranking realistic solutions correctly in a hierarchy is important. If a global civilization is to commit trillions of dollars over the course of many decades to improve the odds of its stable existence, then it should follow the most rational, most economically rewarding, and least environmentally stressful course rather than pursuing inherently inferior alternatives. I believe that the least desirable strategy is to leave the existing excesses, inefficiencies, and irrationalities intact while spending huge sums and creating new environmental problems—some foreseeable and some not. Unfortunately, the monumental unwillingness of both institutions and individuals even to consider eliminating unnecessary conversions and the reluctance to commit vigorously to more efficiency in the remaining ones are now leading toward the embrace of inferior solutions.

Persistent Myths

This book's premise is that myths and misconceptions about energy are nothing new—some that are still with us go back to the nineteenth century—and its purpose is to examine and then debunk several energy myths that are especially cherished today. This should help us to a more realistic understanding of complex energy affairs and introduce some necessarily skeptical perspectives into the often highly uncritical assessments of our future energy options.

Technical Innovation. Obviously, myths and misconceptions are found in any realm of human endeavor. Among recent notable examples, I would include a mistaken belief in an accelerated pace of technical innovation,[8] the expectation of large economic gains from exploiting tropical biodiversity, and the anticipation of a stunning payoff to research in artificial intelligence.

A widespread belief in the acceleration of technical advances owes a great deal to what I call Moore's curse, the idea that the rapid and sustained improvements in the performance of microchips represent the norm in modern inventiveness.[9] In reality, advances in microprocessor abilities are a highly atypical example of technical progress, as I show in chapter 8. A closer examination of tropical biodiversity was to yield a cornucopia of potent new drugs; it has not. And the quest for artificial intelligence has yielded less than astonishing results—the very logic and accomplishments of this decades-long effort are now questioned even by one of the field's creators.[10]

Even a casual observer of the modern energy scene would be aware of exaggerated or failed promises and dreams that did not come true, ranging from the dream of energy self-sufficiency for the United States, first called for by 1973, to the dream of commercially exploited superconductivity to make intercontinental electricity transmission a reality. Those bold expectations got a powerful boost with the discovery of cuprates that superconduct at 30 K, which earned Georg Bednorz and Alex Müller the Nobel Prize in physics in 1987, but more than two decades later, there are no profitable high-temperature superconducting techniques in long-range electricity transmission. Similarly, expectations have been high that the diffusion of cars and buses powered by fuel cells is imminent; in reality, these machines have remained limited to a smallish number of demonstration units built at extravagant cost (see chapter 1).

Some energy myths, including the belief that energy conservation reduces overall energy consumption, are quite venerable, going back more than a century. Others—such as the claims that biofuels derived from crops, their residues (straw, stalks), or wood can displace a large share of liquid transportation fuels refined from crude oil—are more recent. Some attach themselves, barnacle-like, to any substrate. As a result, high-tech worshippers are now telling us that everything will be transformed by nanotechnology, which will, among other things, make possible electricity transmission without losses and incredibly cheap electricity generation by thin-film solar cells, or by genetic engineering, which will create new bacteria from scratch to produce hydrogen or plants to ooze biodiesel.

Sustainability Myths. Other myths are elaborated and vigorously propagated by true believers who seek a world that runs according to their preferences.

Myths involving "sustainability" are now especially popular, although this loaded term means different things to different people, as it is rarely rigorously defined. The list of energy sources and conversion techniques that are to deliver our sustainable energy is therefore rather long and now includes, incongruously, even the fossil fuel industries, which have no intention of vacating the stage they have dominated for more than a century just because sustainability is de rigueur.

Even the producers of electricity generated by coal combustion now claim that their goal is to practice their business in a sustainable fashion, although the concept obviously cannot apply to the resource itself. But I suppose that somebody may yet make a claim regarding the sustainable nature of coal mining; after all, I heard a top executive of one of the world's leading metal mining companies claim that his business is sustainable. In any case, power plant engineers now stress low, or virtually no, environmental impact of new conversions, and they are working on commercializing coal-fired electricity generating plants that sequester carbon and hence produce no greenhouse CO_2 emissions. (I will deconstruct this carbon sequestration myth in chapter 5.) More modestly, various coal gasification and liquefaction processes, captured by the mantra of "clean coal," are promoted as imminent candidates for large-scale commercialization in order to maintain coal's large global role.

Oilmen and the government of Alberta, Canada, boast that the province's oil sands contain more oil than has been found in Saudi Arabia, although most of it would be prohibitively expensive and environmentally ruinous to recover. The Saudis maintain that they can supply the world with enough oil for generations to come, although the state-run company has not offered any verifiable information on the country's actual oil reserves for thirty years. Energy enthusiasts scanning more remote horizons see natural gas hydrates—frozen methane in the Arctic or deep under the sea bottom—as the ultimate fossil fuel whose enormous resources could last for centuries.

A Nuclear Comeback. The nuclear establishment has been trying to stage a global comeback by arguing for the virtue of fission's carbon-free electricity (see chapter 2). It offers new designs of inherently safe reactors and detailed justifications for reviving fast breeders whose experimental designs were abandoned due to excessive cost and safety considerations. And, despite the failed promises of the past half century,[11] promoters of nuclear fusion have never relinquished their hopes that another investment of $20 billion or $30 billion, now through the International Thermonuclear Experimental Reactor (ITER) project under construction in France, will make it the dominant method of energy conversion in a matter of several decades.

Renewable Energy. Much like religious sects that often preach salvation in strictly denominational terms, an army of renewable energy enthusiasts rejects other options and is certain that particular sources or conversions represent the answer to the world's energy problems.

Today's dominant devotions are to wind, particularly in Europe, and crop-derived ethanol, now most fervently propagated in the United States, where scientists seek a fairy tale–like conversion of plant waste into liquid gold, or cellulosic ethanol. (Myths concerning biofuels and wind are discussed in chapters 6 and 7, respectively.) Another large denomination trusts in photovoltaics—the direct conversion of sunlight into electricity; its adherents believe it will soon prevail everywhere, not just in sunny Arizona or Saudi Arabia. Germans were the first taxpayers forced to subsidize heavily what was at the time of its completion the world's largest photovoltaic project—Bavaria's Solarpark, with installed capacity of 10 MW, peak power of 6.3 MW, and an area of 250,000 m^2 divided among three sites[12]—in one of Europe's cloudiest locations. A facility built in a sunny location would produce better results—at least five times that rate would be generated in Sicily or Arizona, for example—but on earth, where the atmosphere interferes and nights follow days, solar conversions are always limited. A superior choice would be to put photovoltaics in the sky as fleets of satellites or, even better, on the moon, with electricity beamed back to Earth by microwaves. For many years, David Criswell, director of the Institute for Space Systems Operations at the University of Houston, has been the leading advocate of this lunar solar power.[13]

I should not forget the devotees of geothermal energy, as well as the minor renewable denominations, including those putting their faith in ocean waves, ocean currents, and ocean thermal differences. The last option involves sinking a long pipe into cold waters (< 4°C, the near-constant temperature of the abyss) beneath the warm subtropical or tropical seas, whose daily high temperatures are > 25°C, and using the temperature difference to generate electricity.[14] Of course, there is that small problem of fundamental thermodynamics: The difference in temperature (ΔT) between the hot and cold reservoir (a mere 20°C) is tiny compared to the difference in a large thermal electricity generating plant, where ΔT is > 500°C. Hence, the efficiency of the process is so low (typically 3–4 percent) that it may take more electricity to pump the deep cold water to the surface than is generated by the process.

But most of these fervent devotions have a common goal: At the end of these renewable rainbows is a near-miraculous, clean, carbon-free hydrogen economy. And before we get there, partial energy salvations will be delivered by hybrid or electric cars (see chapter 1), by compact fluorescent lights or light-emitting diodes, by draconian carbon taxes, by massive sequestration of carbon dioxide, or by stimulation of phytoplankton growth and the subsequent burial of organic carbon in the abyss, a biosequestration method using a natural carbon pump to allow us to emit CO_2 at will.

More Radical Solutions. Other enthusiasts do not have much faith in these means of deliverance and advocate even more radical solutions. Their choice is to proceed with business more or less as usual—that is, to keep on emitting massive volumes of greenhouse gases but to meet the challenge of global warming by turning to geoengineering. Their preferred action would be to play actively with the planet's entire thermal balance by lofting massive amounts of sulfates into the atmosphere to cool the tropospheric temperature. A less intrusive option would be to erect giant maneuverable shades in space to reduce the amount of solar radiation reaching earth.

Such is the energy policy landscape today—a vast collection of myths, misplaced hopes, and uninformed fervor. The purpose of this book is to sort through them, challenge some of the most unrealistic, and refocus on the approaches most likely to be productive. Energy techniques are affected by many external factors, and time is required to determine which are the most valuable and which have the most potential for becoming successfully commercialized innovations.

Challenging the Myths

In part I of this book, I have chosen to look first at a trio of old but remarkably persistent myths which, taken together, illustrate some key attributes of the false notions surrounding energy: their longevity, the attention they receive, and the magnitude of the erroneous anticipations they help generate. Few myths have been around longer than that of a world dominated by electric cars—that belief became commonplace before the end of the nineteenth century. Few have received more attention than the myth of nuclear electricity that will be "too cheap to meter." As for the magnitude of erroneously held convictions, there is perhaps no better illustration than the case of soft, decentralized energy techniques, whose contribution to America's energy supply by the year 2000 was more than 90 percent lower than what their proponents during the late 1970s and the early 1980s expected.

In part II, I devote more space to five current energy issues that have received much coverage by the media, as well as by specialized technical publications: peak oil and the consequences of oil depletion; sequestration of carbon dioxide emitted from the combustion of fossil fuels; fuels from plants, above all America's regrettable, heavily subsidized, and environmentally ruinous production of ethanol from corn; wind-generated electricity, with the Great Plains seen as the Saudi Arabia of wind power; and rapid and fundamental transformations in energy use, exemplified by plans conceived by T. Boone Pickens to reduce America's reliance on oil and by Al Gore to move to a carbon-free mode of generating electricity in a single decade.

In dismantling these energy myths and misconceptions, I necessarily introduce scientific and engineering arguments as well as a good many numbers; these are essential for appreciating the difficulties and magnitudes of the tasks and changes at issue. Bearing in mind that not all readers of the book will be scientists or engineers, I have tried to keep this material as accessible as possible; understanding the book's quantitative explanations calls for nothing more than basic algebra. If a reader is keen to check or replicate the results, a small calculator with scientific notation to make entering large numbers easier would come in handy. The many unit-measure abbreviations are explained in a list at the front of the book.

I have decided to forgo addressing several myths that have already been ably deconstructed elsewhere. That is why I do not deal with the myths of an imminent hydrogen economy or electricity generated by nuclear fusion. Unlike fossil fuels, wind, or solar radiation, which can be extracted from the earth's crust or captured by turbines and photovoltaic (PV) cells, hydrogen is merely an energy carrier and is not found in large amounts underground or in the atmosphere. It has to be produced first by a considerable invest-ment of energy, be it the reduction of methane, the hydrolysis of water, or bacterial metabolism. A study by Olah, Goeppert, and Prakash offers a thor-ough and scientifically impeccable critique of the myths relating to hydro-gen as a fuel.[15] As for generating electricity from nuclear fusion, a brief paper by one of the veterans of its development makes clear how dismal this prospect is, in spite of billions of dollars in funding annually.[16]

With some regrets, I have also decided not to address the notion that energy conservation leads to lower energy consumption. Of course, elimi-nating wasteful energy uses and achieving the highest practical efficiencies in common energy conversions are highly desirable (for reasons ranging from lowered costs to higher comfort), but, ultimately, they do not have the effect for which they are specifically lauded and promoted: In the long run, they do not lower overall energy consumption. The myth that energy effi-ciency reduces energy consumption has been particularly well refuted by Rudin, Herring, and, in a book-length treatment, Polimeni and others.[17] But no recent study can do better than these three sentences from Stanley Jevons's classic treatment of the myth, written 150 years ago:

> **It is wholly a confusion of ideas to suppose that the econom-ical use of fuels is equivalent to a diminished consumption.** The very contrary is the truth. As a rule, new modes of economy will lead to an increase of consumption according to a principle recognized in many parallel instances.[18]

Unfortunately, the mistaken belief Jevons tried to refute has been embraced as a nearly universal truism and has become a matter of such fervent, almost religious, conviction that it appears impervious to any rational critique.

In fact, few myths are easy to dislodge, or even weaken; the tenacious cultivation of myths and the misconceived quest for salvation are as quin-tessential products of human intellect as the effort to understand objective reality and to accept uncomfortable truths. And we can never be sure which

tendency will prevail in any particular case. Still, I hope that this deconstruction of old and new myths will enlighten thoughtful readers, and that
my account of the reluctance to face sobering realities will offer decision-
makers some useful historical perspective. To that end, I conclude the book
by drawing out the specific policy implications of my discussions and by
recapitulating some of their more general lessons.

Finally, an anticipatory concession: Given the number of claims, interpretations, calculations, and figures I present in this book, it is inevitable
that some of them will be questioned and disputed by those who know
more about certain particulars or by those who strongly advocate specific
approaches or solutions. That is as it should be; science is a building under
constant construction that can be made better only by questioning and criticizing and by the contest of numbers and ideas.

PART I

Lessons from the Past

Myths concerning energy have a venerable past. Galileo Galilei (1564–1642) boldly and at considerable personal peril reaffirmed the sun's place at the center of our planetary system, but he thought that heat was a mere illusion of the senses, an outcome of mental alchemies. Francis Bacon (1561–1626), another luminary of early modern science, maintained quite inexplicably that heat could not generate motion, and vice versa. Bacon's contemporaries believed in phlogiston—supposedly the principal component of all combustibles whose liberation (dephlogistication) during combustion leaves behind the residue of calx. This surprisingly durable myth led to many chemical cul-de-sacs before it was finally overturned.

As for more recent history, it is only mildly hyperbolic to assert that current energy policy debates rest mostly on myths. These myths have ranged from a widely held belief that large oil companies have colluded in a decades-long conspiracy to block any viable alternatives to the internal combustion engine (supposedly by buying up or otherwise suppressing relevant patents) to an even more widely held conviction that higher energy efficiency lowers overall energy consumption. Because the number of significant energy myths and misconceptions from the past is embarrassingly large, I limit my focus in the first part of this book to the twentieth century, and specifically to three widespread and persistent myths that have done much to misinform and misguide the public.

This trio of myths and misconceptions includes the promise of electric cars, a myth that began in the 1890s and grew with the commercialization of cars; the promise of nuclear electricity generation in the United States, a post–World War II phenomenon based on the wartime development of nuclear weapons and rushed into commercial uses; and the promise of "soft" energy, or decentralized small-scale energy developments, a myth engendered by great changes in global energy affairs during the 1970s.

Excessive—and so often entirely unquestioning—faith in the efficacy of technical innovations is perfectly illustrated by the century-old vision of the dominance of electric cars. Born at the very end of the nineteenth century, this vision persisted throughout the twentieth and has recently seen a significant renaissance. Once again, the conviction is abroad that the automotive future belongs to electric cars; once again, we read the promise of their astonishing performance, affordable cost, and imminent dominance. Elon Musk, the founder of the latest American electric car company, expects his business to be "a raging success . . . worth multiple billions of dollars."[1] But before getting swept up by this enthusiasm, readers should peruse not only my account of the electric car in chapter 1 below, but also the history of Amory Lovins's Hypercar, briefly described in chapter 3 on soft energy.

My second choice was practically inevitable. In chapter 2, I had to reassess one of the most audacious post–World War II energy myths, whose scope was memorably encapsulated by the claim that nuclear electricity would become too cheap to meter. Who has not heard this boast, and how many people consider it nothing but an apocryphal summation of the hubris displayed by an industry even before it took its first serious steps? In reality, the phrase is the expression of hopes that were at one point held by some of the nuclear industry's best-informed experts. But this myth has a fascinating twist: The failure of nuclear electricity to reach its full potential has not been due to any fundamental technical problems or economic considerations. Instead, the decisive factors have been a faulty perception of risk and of the changing nature of electricity markets, leaving many observers to consider its comeback not only possible but desirable, even inevitable.

And there could hardly be any more fitting contrast to the vastly unrealistic expectations raised by the proponents of nuclear fission than the kindred yet diametrically opposite belief in soft energy, discussed in chapter 3. The genesis of this belief came from Amory Lovins's uncompromising opposition to anything nuclear in particular and to anything large-scale in general. Lovins's conviction that soft energies—conversions of the biosphere's renewable energy flows carried out in small, decentralized facilities—would secure modern civilization's energy future turned out to be as false as it was appealing.

1

The Future Belongs to Electric Cars

The myth that the future belongs to electric vehicles is one of the original misconceptions of the modern energy era, going back to the very introduction of the first practical passenger cars. During the new industry's first two decades, many engineers and observers were not certain which type of machine—steam, electricity, or gasoline powered—would eventually dominate the market.[1] Four generations of experience with high-pressure steam engines made possible the building of some remarkably powerful, nimble, and fast steam-powered vehicles. In the very first car race, in July 1894, vehicles with Daimler and Maybach's gasoline engine took four out of the first five places—beaten by a steam-powered De Dion and Bouton machine. In 1902 in Nice, Leon Serpollet's beak-shaped, steam-powered racer set a new speed record for one kilometer in 29.8 seconds (equivalent to 120.8 km/h), and in 1906, Francis and Freelan Stanley's steam car ran the fastest mile at a speed equivalent to 205.4 km/h.[2]

During this period, electric cars appeared to have even greater promise because their impressive performance was easier to reproduce in reliably operating commercial vehicles. Unlike the drivers of steam cars, drivers of electric cars did not have to reckon with high-pressure boilers and escaping steam. Unlike gasoline-powered cars in this time before the electric starter and automatic pump, electric cars did not require dangerous, arm-wrenching cranking, awkward refills with a highly flammable liquid, or strenuous gear shifts, nor was there any fuel smell from a largely exposed engine. In 1896, a Riker electric car won the first U.S. track race at Narragansett Park in Rhode Island when it decisively defeated Frank Duryea's gasoline vehicle. On April 29, 1899, the bullet-shaped electric La Jamais Contente, driven by Camille Jenatzy, broke the 100 km/h barrier by briefly going at 105.88 km/h.[3]

Meanwhile, the commercial introduction of electric cars began in 1897 with a dozen Electric Carriage and Wagon Company taxicabs in New York. In 1899, U.S. carmakers produced more than 1,500 electric vehicles, compared to just 936 gasoline-powered cars.[4] In 1901, Pope's Electric Vehicle Company was both the largest manufacturer and the largest owner and operator of motor vehicles in the country.[5] Other well-known makers included Anthony Electric, Baker Electric, Detroit Electric, and Studebaker. The diffusion of electric cars led to the emergence of a new infrastructure aimed at overcoming the limited range of these vehicles. By 1901, it was possible to travel by electric car from New York to Philadelphia, thanks to six charging stations that were built in New Jersey; and by 1903, Boston had thirty-six charging sites.[6]

Electric- versus Gasoline-Powered Cars

Few people believed more strongly in the eventual dominance of electric cars than Thomas Edison, the inventor of the modern electric system. This conviction brought about one of the most consequential partings in the history of technology. Henry Ford was hired as the chief engineer at Detroit Edison Illuminating Company, but he continued his experimental work on gasoline engines. Executives of the Edison Company objected to this work and offered him a promotion to general superintendent, in his words, "on the condition that I would give up my gas engine and devote myself to something really useful."[7]

Edison's famous stubbornness led him to predict that electric cars would eventually prevail, even after the Electric Vehicle Company went bankrupt in 1907 and Ford launched his affordable and reliable gasoline-powered Model T in 1908. Edison spent almost the entire first decade of the twentieth century trying to develop a high-density battery that could compete with gasoline.[8] The result of this costly effort, introduced in 1909, was Edison's nickel-iron-alkaline battery, which came to be used mainly as a dependable standby source of electricity rather than a competitive prime mover for vehicles.

During the next fifteen years, improvements in gasoline engines and advances in car construction combined to make electric vehicles the losers in the vehicular evolutionary race.[9] The relevant innovations included the universal adoption of assembly-line car manufacturing, introduced with Ford's Model T in 1908; the electric starter, which eliminated cranking, patented by Charles Kettering in 1911 and introduced in Cadillacs in 1913; and Thomas Midgley's solution to engine knock—the addition of tetraethyl lead to gasoline, beginning in 1924. Interest in electric propulsion never faded among small groups of engineering enthusiasts, but by the 1930s there were no commercial makers of electric cars.

Recent History of Electric Cars

After World War II, abundant and cheap gasoline coupled with affordable mass-produced cars left no room for electrics. Henney Coachworks and National Union Electric Company joined to make the first transistorized electric car, Henney Kilowatt, in 1958, but they abandoned its production three years later after selling fewer than a hundred cars. Even the OPEC-driven oil price rises of 1973–74 and 1979–81 were not enough to resurrect a commercial commitment to electric cars. To be sure, a tiny, angular, and ugly CitiCar, made by Sebring-Vanguard and looking much like an inverted wheelbarrow, was introduced in 1974 at the Electric Vehicle Symposium in Washington, D.C., and in 1975, the U.S. Postal Service bought 350 electric jeeps from AMC (American Motors Corporation). But neither of these moves had any long-lasting consequences, and this tentative interest had disappeared by 1986, after OPEC's extortionary oil prices collapsed.

What looked like the most promising beginning of a real electric car resurrection came exactly a century after the vehicles began to enjoy their first success, and it was thanks to efforts to improve the quality of California's notoriously polluted air. In 1990, the California Energy Commission mandated that by 1998, 2 percent of all new vehicles (about 22,000 cars) sold in the state would have to be electric, and that by 2003 the share of zero-emissions vehicles—presumably mostly electrics—should reach 10 percent of the state's car sales, or close to 150,000.[10] But subsequently these requirements were greatly weakened, and none of the original goals were achieved; no truly commercial electric cars became available during the 1990s (Kirsch 2000).

In 2001, the California Air Resources Board redefined the goal for the year 2003: At least 10 percent of newly sold vehicles were to have low emissions, but only 2 percent were to have zero emissions.[11] A year later, the now defunct DaimlerChrysler joined General Motors in a lawsuit against the California Air Resources Board asking that it repeal all zero-emissions vehicle mandates, and in 2003 GM decided to stop leasing its EV 1, a two-seat sports car powered by lead-acid batteries and produced in small numbers since 1996. This prototype was discontinued by the end of 2004 amid conspiratorial accusations that the usual suspects (large car and oil companies) wished it to fail.[12] But Kirsch put it best: Electric cars have never been a replacement for America's family sedans, nor will they now replace vans and SUVs; they are a niche product whose small market has always translated into small profits and, hence, into no more than a very reluctant embrace by major car companies.[13]

Other electric vehicles were introduced during the late 1990s, but in all cases their production lasted only a few years. Those with nickel-metal hydride batteries included GM's Chevrolet S-10 pickup (1997–98, only about 50 sold), Honda's EV Plus sedan (1997–99, some 300 sold), Toyota's RAV4 SUV (1997–2002, about 1,200 sold), and Chrysler's EPIC minivan (introduced in 1997, 80 sold), while Nissan's Altra station wagon (produced between 1998 and 2000, with fewer than 150 vehicles sold) had a lithium-ion (Li-ion) battery. The high cost of these automobiles was an expected drawback, but the companies did not even try to reduce it; instead, they became infatuated with hybrid vehicles, which were seen as the most likely choice for future passenger-car propulsion. Once again, purely electric cars failed to make even a modest commercial comeback.

Recent Electric Models

The myth of dominance by electric cars, however, refuses to die. It turns out that even the electric hybrids, now offered by all major carmakers, have not entirely eliminated the dream of the purely electric car. Enthusiasts still await its ascendancy, and media reports continue to proclaim it the next decade's choice. In the United States this most recent wave began with the tiny Tango, the expensive Tesla Roadster, and the supposedly "game-changing" GM Volt, all of them touted as finally ushering in the era of the electric car's triumphal ascendance.

Tango was originally a Smart Car—a Mercedes-made, gasoline-powered two-seater whose city performance was rated at 4.6 L/100 km, or just over 50 mpg—converted to an all-electric drive by Hybrid Technologies (2007) in Nevada. And since we all know that people who care most about the planet live in Hollywood, nothing could make us take this vehicle more seriously than the fact that George Clooney bought it "sight unseen" and then graciously consented to be photographed while leaning languidly against its shiny minibody: "Clooney's Tango! WoW!!!"[14] The first version carried 218 kg of lithium-ion batteries and needed six to eight hours of charging; its range was advertised as 193–241 km and its top speed was 128 km/h.

By 2009, Hybrid Technologies had morphed into EV Innovations[15] and offered not only the tiny car, now renamed Dash, but also lithium-battery versions of the PT Cruiser and Morris Mini Cooper (each for more than $50,000), as well as a sports car, a motorcycle, and a moped. Meanwhile, Tango became the name of an ultra-narrow (99 cm), easy-to-park, all-electric car, with about 900 kg of standard lead-acid batteries. Its makers, the Commuter Cars Corporation of Spokane, Washington, call it a "revolutionary commuter vehicle."[16] But its price may stand in the way of the car's revolutionary function. In 2009, Tango T600 cost $108,000 and there was still no mass production of a more affordable T200 model.

Tesla Roadster has an even more newsworthy pedigree—media cannot stop gushing over the fact that it comes from Silicon Valley, although it really does not. Elon Musk, the founder of PayPal, set up Tesla Motors partly with his own money to produce a powerful electric car that could, according to a wide-eyed *Vanity Fair* writer, "spell the end of the internal-combustion engine."[17] The Tesla Motors website[18] promised much more: a green future and "a peaceful solution to oil wars" through the introduction of "gasoline-free" cars.[19] This is not a joke, but a quotation, and it places a very heavy burden on a frivolous machine whose retail price began at $92,000, was raised in the spring of 2007 to $98,000, and by the fall of 2008 to $109,000. In 2009 (with $7,500 of U.S. federal tax credit), it sold for $101,500. The car went into regular production in March 2008, and some 900 cars were delivered by the end of 2009. You can have a Roadster, too, if you put down an initial fee of $5,000, followed by $55,000 to lock in your production spot—and wait.

My advice: Do not be surprised if the end of gasoline cars and the emergence of electric vehicle supremacy do not unfold exactly along the lines anticipated by Elon Musk and *Vanity Fair*. The Roadster is essentially an extended-wheelbase British Lotus Elise loaded with 6,831 lithium-ion batteries.[20] The energy density of these batteries can be as high as 160 Wh/kg, which is four times that of standard lead-acid cells, and they would give the car a range of 400 km and enable it to be recharged in less than four hours. The price makes it abundantly clear that the market for this "high-performance" two-seat "sports" car is a smallish group of showoffs and I-already-have-everything-else customers.

These buyers are smitten by the fact that it can reach more than 200 km/h and accelerate to 100 km/h in less than four seconds, and that it can pin a driver against the back of the seat like in a fighter plane.[21] The Roadster's pricing and appeal would have been familiar to the promoters of a similar class of cars before World War I, making it anything but a "reinvented" car. Because the list of its key promoters and "founding owners" includes assorted Silicon Valley executives (leaving aside celebrities like Clooney), many people have begun to assume that in such techno-savvy hands the electric car is now bound to follow the trajectory of personal computers or mobile phones. They will be badly disappointed. And so will be the owners: They might take off fast, but they will learn about the Roadster's real range when it is driven as a sports car, and about its deteriorating performance with time.[22]

Electric Cars and the Supply of Electricity

As the third wave of high crude oil prices washed over the world between 2005 and 2008—a wave that coincided with fear of an imminent peak in worldwide oil extraction and rising concerns about America's addiction to imported oil—the electric car acquired more than a niche appeal. Its main selling point is an ability to run without any imported fuels. In 2007, this grand benefit was featured by *Foreign Policy* as one of "21 Solutions to Save the World": "Flip switch," the article read "Nearly all the world's oil will soon be in the hands of unreliable autocrats. It's time we went electric."[23]

Since that time there have been so many announcements of new electric car projects that their up-to-date lists could be maintained only online. The leading U.S. contender is the four-seat Chevrolet Volt, a car that is powered primarily by an electric motor (rechargeable from a standard outlet) and that uses a small (one liter) fuel engine only as a generator to extend the driving range when the battery storage gets low.[24] Before its 2009 bankruptcy, GM saw the vehicle as a fundamental component of its long overdue strategy to regain its market share and reinvent itself as a competitive car company; after the emergence from its bankruptcy, GM sees Volt as a car that will reestablish its technical competence, and it hopes that the vehicle's introduction in 2010 will put the company ahead of its competition.

The production Volt, a little larger than a Honda Civic, does not look like the model unveiled in 2007, although it retains "a similar set of visual cues and some of the features that were on the concept car."[25] Chevrolet's plan is to build 10,000 units in 2010 and 60,000 units in 2011 (priced at about $40,000 per car), and even if the post-2011 production rises by a (most unlikely) rate of 50 percent per year (compounded), there would be about 2.3 million Volts on the road by 2020, amounting to less than 1 percent of all U.S. vehicles. Volt may be a revolutionary gamble for GM, but even its best imaginable success will not transform America's car fleet in a hurry.

Other notable American entries include Tesla's Model S, to be available in 2011 at little more than half the price of the Roadster; an expensive ($88,000) Fisker Karma (working on the same principle as the Volt); and Coda, a more affordable ($35,000) midsize sedan whose modest mission is to end our dependence on oil. There are, of course, new Japanese, and now also Chinese, entries, but the Renault-Nissan alliance has the boldest commitment to an all-electric future: It plans to lease a small array of battery-powered machines because its chairman, Carlos Ghosn, believes that electrics will command 10 percent of the global car market by 2020.[26] And Shai Agassi's Better Place car company plans to sell Renault-Nissan cars but to own their batteries and build extensive networks of recharging stations, first in Denmark and Israel.[27]

Such visions need so many reality checks that I will list just a few of the key ones. With the global car demand forecast at more than 80 million vehicles a year by 2020, carmakers would have to boost their production of pure electrics to more than 8 million in just one decade to make Renault's forecast a reality. How likely is that, given the fact that hybrid cars, which have been around for more than a decade, claimed less than 3 percent of the U.S. market in 2009? How readily will the requisite tens of millions of batteries be available when manufacturers are quick to unveil new, bold electric car plans but slow to commit to massive battery orders?[28] And how will the car owners in large cities, where 30–60 percent of all cars are parked curbside, charge their vehicles?

Obviously, mass construction of a fairly high density of charging stations must precede any mass ownership of pure electrics outside of the suburbs, where the vehicles could be charged in their garages. That is why researchers at IHS Global Insight put the share of pure electrics at just 0.6 percent of world sales in 2020,[29] why most published scenarios put the likely share of electrics at no more than 25 percent of new sales by 2050,[30] and why even Germany, ready to subsidize the ownership of electric cars with major new incentives starting in 2012, is aiming at putting no more than about 1 million electric vehicles on its roads by 2020.[31] With nearly 55 million vehicles total on German roads in 2010, electrics would claim less than 2 percent of all German passenger cars by 2020. Any beliefs in an imminent massive takeover of the global car market by pure electrics are thus highly unrealistic.

But even if electrics were to do better than can be realistically expected, we still have to look at what flipping the switch would do to actual energy demand. Only simple algebra and a string of realistic assumptions, based on the typical performance of electric cars and on the latest transportation statistics, are needed to calculate what this would mean in America's case. In calculating the overall burden that an entirely or partly electric fleet would put on the country's electricity supply, we would be naïve to assume that either converted Smarts or PT Cruisers, and even less so $100,000 Roadsters, will be America's choice for daily transportation. An electric version of a car whose size would correspond to today's typical American vehicle (a composite of passenger cars, SUVs, vans, and light trucks) would require at least 150 Wh/km; and the distance of 20,000 km driven annually by an average vehicle would translate to 3 MWh of electricity consumption.

In 2010, the United States had about 245 million passenger cars, SUVs, vans, and light trucks; hence, an all-electric fleet would call for a theoretical minimum of about 750 TWh/year. This approximation allows for the rather heroic assumption that all-electric vehicles could be routinely used for long journeys, including one-way commutes of more than 100 km. And the theoretical total of 3 MWh/car (or 750 TWh/year) needs several adjustments to make it more realistic. The charging and recharging cycle of the Li-ion batteries is about 85 percent efficient,[32] and about 10 percent must be subtracted for self-discharge losses; consequently, the actual need would be close to 4 MWh/car, or about 980 TWh of electricity per year. This is a very conservative calculation, as the overall demand of a midsize electric vehicle would be more likely around 300 Wh/km or 6 MW/year.[33] But even this conservative total would be equivalent to roughly 25 percent of the U.S. electricity generation in 2008, and the country's utilities needed fifteen years (1993–2008) to add this amount of new production.[34] As this power for electric cars would have to come on top of the demand growth by households, services, and industries, it would be exceedingly optimistic to expect such an increment could be in place in less than twenty years, even if the availability of requisite resources and conversion techniques needed to generate this electricity were ensured.

But Kintner-Meyer, Schneider, and Pratt (2007) would say that these calculations of additional electricity generating capacity are largely incorrect, because as of 2001 (the baseline of their calculations), the United States could produce enough electricity from its underutilized capacity to power about 73 percent of all light-duty vehicles on the road at that time, that is about 173 million cars, pickups, and SUVs. Their calculation assumes plug-in hybrid electric vehicles with batteries able to satisfy average drives of 33 miles (or about 53 km) a day, with all the additional generation coming only from coal- and natural gas–fired power plants. It would call for increasing the average capacity factors of those plants from, respectively, about 73 percent and 40 percent to as much as 85 percent, and for recharging the cars with electricity produced in excess of the existing average load at all hours; if the charging periods were only between 6 p.m. and 6 a.m., the additional electricity generated without adding any new capacity would power not 73 percent, but only about 43 percent, of all light-duty vehicles.[35]

These conclusions have been welcomed in the blogosphere under such headlines as "Cost of Converting Entire U.S. to Electric Cars? Zero." That is, of course, a ridiculous misinterpretation; Kintner-Meyer and his colleagues make no such claims. At the same time, theirs is a purely theoretical exercise, and they acknowledge that whether the existing generation infrastructure and capacity mix could continue to support such high loads is questionable. Moreover, their "valley-filling" approach—using spare capacity during hours of low electricity demand—would necessitate that the recharging of tens of millions of vehicles be managed nearly perfectly—that is, without causing new peaks of electricity demand. That feat would require a great deal of new infrastructure, with multitudes of recharging stations at workplaces and in parking lots to provide electricity also during daytime hours away from home, and unprecedented levels of coordination and automation.

But even if—highly improbably—no new capacity were needed, additional electricity would still have to be generated. In 2008, 49 percent of America's electricity was generated by coal combustion, 20 percent by natural gas, about 20 percent by nuclear fission, and 6 percent by hydroelectricity; the rest was produced from fuel oil, wind, and geothermal energy.[36] The average source-to-outlet efficiency of U.S. electricity generation is about 40 percent and, adding 10 percent for internal power plant consumption and transmission losses, this means that 11 MWh (nearly 40 GJ) of primary energy would be needed to generate electricity for a car with an average annual consumption of about 4 MWh.

This would translate to 2 MJ for every kilometer of travel, a performance equivalent to about 38 mpg (6.25 L/100 km)—a rate much lower than that offered by scores of new pure gasoline-engine car models, and inferior to advanced hybrid drive designs or to new DiesOtto engines (described below). A very similar result is obtained using the assumptions spelled out by Kintner-Meyer and others. This means that an all-electric national fleet would offer no overall primary energy savings and no carbon emissions advantage when compared to the alternatives of a highly efficient gasoline-car fleet or the large-scale adoption of hybrid vehicles—unless, of course, all of the electricity consumed by the all-electric vehicles were generated by renewable conversions rather than by the current mix of generation relying on coal, natural gas, nuclear fission, and water power.

The latest European report on electric cars—appropriately entitled *How to Avoid an Electric Shock*—offers analogical conclusions.[37] A complete shift to electric vehicles would require a 15 percent increase in the European Union's electricity consumption, and electric cars would not reduce CO_2 emissions unless all that new electricity came from renewable sources:

"Electric cars powered by wind or solar energy are obviously superior. But if the electricity comes from coal, hybrids perform better."[38] In global terms, the International Energy Agency calculated that the total transport electricity demand would become 20 percent of the total generation by 2050, and would require more than 2 TW of additional capacity.[39] That would mean increasing the current global capacity by about half and adding that amount entirely in renewable conversions—a plausible achievement after fifty years, but an impossible task during the coming generation (twenty to twenty-five years).

More Efficient Gasoline Engines

We also have to keep in mind that gasoline-fueled internal combustion engines can be made considerably more efficient. By far the most promising recent development is the ingenious DiesOtto design by the Daimler engine research laboratories.[40] The new engine starts by spark ignition of directly injected gasoline and runs in this conventional (Otto) mode at full load, but during partial load (that is, at the low and medium speeds most common in typical driving), it morphs into self-igniting (diesel) operation; in addition, the engine is supercharged. This combination means that an engine with fewer cylinders and a smaller displacement offers great performance, low fuel consumption, and very low nitrogen emissions.

Daimler's goal of making the gasoline engine as economical as a diesel is thus well within reach, and it is quite realistic to expect the future DiesOtto machines to achieve more than 60 mpg, above the rate for today's highest-rated hybrid car (the Toyota Prius is rated at 55 mpg for the combined city–highway cycle). In addition to variable valve control, innovations currently under development include engines that switch between four- and two-stroke modes and can even shut down a cylinder or two during spells of low power demand, operating temporarily on an Atkinson (ultra-lean combustion) cycle rather than on the standard Otto cycle.

The only way electric cars could reduce global carbon emissions would be if all the additional electricity needed to power them came from carbon-free energies. Here, again, the considerations of scale come into play. Inherently low load factors of wind or solar generation, typically around 25 percent, mean that adding nearly 1 PWh of renewable electricity generation would require installing about 450 GW in wind turbines and PV cells, an equivalent of nearly half of the total U.S. capability in 2007. That would obviously be an enormous undertaking whose accomplishment would span decades (in the United States, it took nearly three decades to add that capacity) and whose cost would be at least half a trillion dollars.

Better batteries and better car performance would obviously reduce the aggregate electricity demand. Announcements of new breakthroughs in battery design—claiming unprecedented power densities and durabilities—have become increasingly common, but I have not seen any proofs that the transformation of most of these new designs into commercial vehicle-powering units sold by the millions every year will make cars much better than the best-performing models of today. My best guess is that something like Edison's frustrating experience lies ahead, and not any Intel-like rate of advance.

Cars with the lighter yet more durable bodies that are better suited to electric drive are easier to introduce than any new superior batteries, but even that innovation would require at least two average ownership cycles (the U.S. mean is now about nine years) before such vehicles could constitute a significant share—say, at least 15 percent—of the total fleet. All of these realities suggest that a meaningful transition to electric drive (as opposed to a token presence of expensive cars) would represent an enormous technical and investment task that could be accomplished only very gradually. And even if we assume a doubled electric drive efficiency, other considerations would have to be taken into account if an entirely or even largely electric fleet were to run on renewable electricity. Above all, there would be a great mismatch between the peak wind and solar capacity during the day and a huge nightly demand spike as tens of millions of vehicles were recharging in their home garages.

I have not even mentioned the not-so-ideal properties of the Li-ion batteries, which are now so highly touted.[41] These devices lose the charge even when they are idle. Hence, an aging battery, even one that has never been used, will have a more limited life span than a new one. The normal expectation is for two to three years of service, while major car components are now designed to last at least ten years.

Tesla engineers expect the power of the car's battery pack to degrade by as much as 30 percent in five years, but this irreversible capacity loss is temperature dependent—at the freezing point and at 100 percent charge, degradation is about 6 percent after one year; at 25°C it is 20 percent; and at 40°C it is 35 percent. This is not a negligible consideration for all the vehicles driven through the American summers and, especially, for the roughly 40 percent of them operated in the Sunbelt, where summer temperatures of 30°C or higher are common.

Obviously, smaller vehicles and better batteries (with everybody driving a tiny Tango) would lessen the required electric load and would help move electric cars from a niche market into the mainstream. But to believe that flipping the switch and going electric will solve America's automotive dependence on imported oil, either in the near or long term, is utterly unrealistic, even delusionary.

Even when leaving aside the profound—and in the final analysis highly uncertain—consequences of the car market collapse of 2008–9 and the bankruptcies of two out of three of the major U.S. automakers, it is important to stress that unexpected developments may change the outlook. This is the case with all forecasts of new techniques in their earliest stages of commercialization, and it is possible that a shift to electric cars will proceed faster than seems likely in 2010, and that this shift will prove part of the remedy to America's automotive excesses and failures. But even then it is unlikely that it will be the dominant factor of automotive innovation during the second decade of the twenty-first century; in any case, it will be decades, rather than years, before we can judge to what extent electric cars offer a real substitute for vehicles powered by internal combustion engines and contribute to more efficient personal transportation in the United States.

2

Nuclear Electricity Will Be Too Cheap to Meter

"Too cheap to meter" is perhaps the best known, and certainly the most quoted, statement concerning the future of a new energy conversion—and it is not apocryphal. In 1954, Lewis L. Strauss, chairman of the U.S. Atomic Energy Commission from 1953 to 1958, told the National Association of Science Writers in New York that our children will enjoy in their homes electrical energy too cheap to meter. It is not too much to expect, he said, that our children will know of great periodic regional famines in the world only as matters of history, will travel effortlessly over the seas and under them and through the air with a minimum of danger and at great speeds, and will experience a lifespan far longer than ours, as disease yields to medicine and man comes to understand what causes him to age. This is the forecast of an age of peace.[1]

It was clearly a generic futuristic vision, and Strauss did not even make an explicit link between *nuclear fission* and electricity too cheap to meter— a fact that led some to argue later that he had in mind the commercialization of *nuclear fusion*. (Commercial fission uses neutrons to split nuclei of the heaviest natural element, uranium; fusion, the joining of the lightest nuclei, powers stars and hydrogen bombs, and its commercialization remains as elusive as ever.) And his statement should obviously not be taken literally, as even "free" electricity would have to be transformed, transmitted, and distributed to users, requiring the construction and maintenance of an extensive and costly infrastructure. Perhaps the most logical explanation of the statement is that he wished to suggest electricity would be so cheap that households could be charged a fixed monthly or annual rate rather than pay for the amount actually consumed.

But all these qualifications are irrelevant. Strauss's phrase acquired a life of its own, as it came to embody technical hubris—the unrealistically boastful attitude of arrogant innovators—and as it has been used by the critics of nuclear power to disparage the industry's credibility ever since. The reality that surrounded that unfortunately hyperbolic statement was, however, more complex. Questioning the technique's maturity, costs, and potential risks, many power engineers and utility economists were not at all enthusiastic about the push to develop nuclear generation and raised doubts about whether it was even needed in a modern electricity supply. This group included the first chairman of the U.S. Atomic Energy Commission, David E. Lilienthal. In 1955, during the first International Conference on the Peaceful Use of Atomic Energy, Lilienthal wrote in his private journal that the recent history of nuclear development "is characterized more by salesmanship, propaganda, and overzealousness than sense. These men are fanatics or zealots; caveat zealot!"[2] But these damning words, unlike Strauss's vision, have not become part of nuclear lore.

The zealots, including many prominent scientists and engineers, clearly shared Strauss's vision, and this brand of enthusiasm persevered into the 1970s. Consequently, I take Strauss's phrase as an embodiment of the high hopes with which many knowledgeable people invested nascent nuclear generation. The realities have been very different: in spite of some undeniable successes, nuclear power has fulfilled only a fraction of its original promise. More important, nuclear fission had eventually ceased to be a major energy supply option in virtually all advanced economies. A look at why such high hopes were held for nuclear power to begin with, and why these hopes were never fulfilled, offers one of the most fascinating examples of a sudden reversal in the history of technical innovation.

Peaceful Uses of Nuclear Fission

During the late 1940s, in the aftermath of the Manhattan Project and the bombing of Hiroshima and Nagasaki, there was no demand for nuclear electricity generation and no economically compelling reason to develop it rapidly. American utilities and makers of power plant equipment assessed the prospects of nuclear power rather cautiously, as did many notable nuclear scientists. All the same, the post-1950 development of nuclear generation proceeded fairly quickly, and Rebecca Lowen suggests a number of reasons why this was the case.[3] An important consideration for the United States was certainly eagerness to ease the sense of guilt over the bombing of Hiroshima and Nagasaki by demonstrating a peaceful use of fission. The political and strategic considerations—the desire not to be bested by the Soviet Union or, for that matter, by the United Kingdom and Canada, two Western countries that formulated their own early nuclear programs—were no less important.

In December 1953, after the explosion of the first Soviet thermonuclear device, President Dwight D. Eisenhower announced his Atoms for Peace plan. The plan sought to demonstrate America's nonthreatening, peaceful nuclear capability and was also designed to attract the attention of nonaligned countries interested in new forms of energy—the economical generation of electricity for domestic use was not its primary goal. The plan required an operational reactor, and the only reactor design available at that time for uses other than weapons production was the one used to power new U.S. submarines. The first nuclear-powered submarine, *Nautilus*, was launched in January 1954 following rapid innovation efforts led by Admiral Hyman G. Rickover.[4] That is why the U.S. Atomic Energy Commission assigned the country's first civilian nuclear power project to Rickover, who used the same kind of General Electric pressurized water reactor (PWR) for the Duquesne Light station in Shippingport, Pennsylvania, that he used in his submarines. The reactor became critical on December 2, 1957, more than a year after the first British plant began operating at Calder Hall.[5] The U.S. transfer of a submarine reactor design to electricity generation established the PWRs as the industry's globally dominant technique—but it also rushed deployment of what was considered at that time an interim reactor design, and the resulting technical lock-in had far-reaching consequences.[6]

The launch of the first national nuclear program was not followed by any rush into nuclear generation during the late 1950s, and U.S. utilities ordered only 12 reactors before 1965. The real commercial breakthrough came during the late 1960s, with 83 orders between 1965 and 1969, and by the late summer of 1970 the United States had 107 units on line, under construction, or past the point of reactor purchase. Nuclear generation was clearly taking off, and expectations for its eventual impact were rising even faster.

In 1971, Glenn Seaborg, chairman of the U.S. Atomic Energy Commission and a Nobel Prize–winning chemist, delivered an address at the fourth International Conference on the Peaceful Uses of Atomic Energy that was even more visionary than Strauss's 1954 talk.[7] By the year 2000, Seaborg said, nuclear energy's "unimagined benefits" were to improve the quality of life for most of the world's population. Fission reactors would not only generate nearly all of the world's electricity; they would also transform the world's agriculture by energizing food production complexes. Here Seaborg was promoting the large nuclear complexes ("nuplexes") first proposed by Richard L. Meier in 1956 and later elaborated by the Oak Ridge National Laboratory.[8] These complexes, centered on large nuclear plants and located in coastal desert areas, were to produce energy for the desalinization of sea water, synthesis of fertilizers, industrial activities, and intensive crop cultivation that would make deserts habitable.

Many other nuclear wonders were to be in place by the year 2000: "Giant earth-stationary satellites bearing compact nuclear reactors will broadcast television programs"; nuclear-powered tankers and other merchant ships "will almost certainly ply the seas"; "peaceful nuclear explosives will be employed on a widespread scale" in underground mineral mining and used to modify the earth's surface, alter river flows, and construct new canals and new harbors in Alaska and Siberia; and "nuclear propulsion" would carry men to Mars.[9] With physicist William Corliss, Seaborg advocated the creation of underground cities—a "nether frontier"—that would be carved out using nuclear explosives. The surface could then be returned to wilderness, and visiting it would be just a matter of getting into an elevator.[10] The implication was that without nuclear energy, civilization would slowly grind to a halt.

But Seaborg's enthusiasm was not an idiosyncratic oddity. Other experts may have been somewhat less visionary, but they foresaw a similar course ahead. David J. Rose, a leading nuclear energy expert at the Massachusetts Institute of Technology, observed that the oil price rises of 1973–74 "reinforced what seemed evident even several years ago: a massive switch to nuclear power for electricity generation, and perhaps later for other purposes." He considered predictions of 1 TW of installed nuclear power by the year 2000 plausible, and he called nuclear power "the largest coherent technological plunge to date, with long-lasting consequences."[11] Hans

Bethe, winner of the 1967 Nobel Prize in physics for his work on nuclear energy production in stars, argued that "the vigorous development of nuclear power is not a matter of choice, but of necessity."[12]

During the early 1970s, encouraging news regarding the construction costs of American nuclear reactors seemed to justify an optimistic outlook. Granted, the six nuclear generating units completed between 1970 and 1974 by Commonwealth Edison of Chicago cost $147–$280/kW$_e$, compared to $113–$218/kW$_e$ for the utility's coal-fired units that were put in service between 1965 and 1975. But this comparison was deceptive. Higher fuel costs and the need to install flue gas desulfurization (FGD) scrubbers in the coal-burning plants to remove sulfur dioxide from emissions meant that the operating cost of nuclear generation was nearly 20 percent lower than that for units with FGD, and the same as that of units without scrubbers. Rossin and Rieck thus concluded that "nuclear plants have been good investments and have produced substantial savings to consumers."[13] And it would have been logical to expect that the first round of OPEC-driven high crude oil prices in 1973–74 would only strengthen the case for nuclear power, as this option offered an attractive way to reduce dependence on expensive and increasingly uncertain supplies from the Middle East.

Retreat from Nuclear Power

In reality, the quintupling of world oil prices had a decidedly negative effect on nuclear fortunes. Higher oil prices, higher inflation, lower rates of economic growth, and a belated effort to conserve electricity were key factors that helped reverse the decades-long era of high annual growth in electricity demand. Until 1970, this demand had doubled roughly every ten years, with average annual growth of around 7 percent; but after 1973 the growth rates dropped to just 2–3 percent per year, and in some nations and regions they entered a prolonged period of stagnation. The United States went from capacity shortages during the early 1970s to a large capacity surplus during the 1980s.

And this was only one of the major factors that halted, and then reversed, the nuclear expansion. Large numbers of new power plant orders placed during the early 1970s led to shortages of skilled labor and to major delays in construction, problems made much worse by increasingly cumbersome

FIGURE 2-1

TRENDS IN COST ESTIMATES FOR AMERICA'S 1 GW$_e$ NUCLEAR
GENERATING UNITS

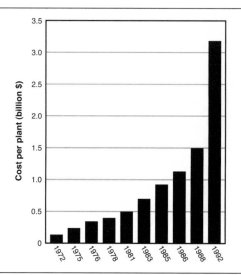

SOURCE: Based on Olds (1982).
NOTE: Graph shows projects with starting dates between 1967 and 1980.

regulations. Long delays and huge cost overruns became the norm. In the early 1970s, it took about 50 months to complete a new nuclear generating unit, whereas in the early 1980s it took 130 months; during that period, requirements for skilled onsite work increased by more than 13 percent a year. As of January 1, 1971, the United States had some hundred codes and standards applicable to nuclear plant design and construction; by 1975, the number had surpassed 1,600; and by 1978, 1.3 new regulatory or statutory requirements, on average, were being imposed on the nuclear industry every working day.[14]

As a result, unit costs began to escalate (see figure 2-1). A plant whose construction began in 1980 for completion in 1992 was expected to cost well over $3,000/kW$_e$, whereas a unit completed in 1975 after less than six years of construction cost just $240/kW$_e$. And worse was to come: Diablo Canyon plant in California, whose original projected cost was $450 million, cost $4.4 billion, while New York's Shoreham, nine years behind schedule, cost $6 billion rather

FIGURE 2-2

THE WORLD'S NUCLEAR GENERATING CAPACITY, FORECAST AND ACTUAL

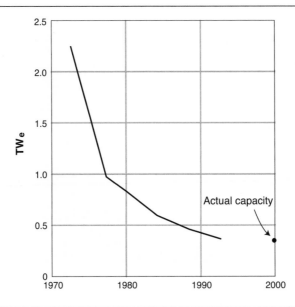

SOURCE: Based on a graph in Semenov and Oi (1993, 4).

than the projected $241 million. By the late 1980s, the cost overruns became so bad that a detailed study of nuclear plants under construction concluded that, with the possible exception of those in the Southeast, the least expensive choice was not to complete them—and many were not completed.[15]

No new nuclear plant was ordered in the United States after 1978, and all thirteen plants ordered between 1974 and 1978 were eventually canceled. By the mid-1980s, it was clear that the reality of nuclear power would bear little resemblance to the original vision. During the early 1970s, the highest estimate of global nuclear capacity available by the year 2000 reached 4 TW, while the International Atomic Energy Agency expected that the world would have as much as 2.5 TW_e installed in fission reactors. By 1980, the expected total for the year 2000 dropped to less than 1 TW, and by 1990 it stood under 500 GW, less than a fifth of the original forecast (figure 2-2).[16] The actual 2000 total, 351 GW_e in 438 operating stations, was lower still.[17]

Hope for Fast Breeder Reactors

Even as this retreat was unfolding, however, a new hope was arising. A worldwide expert consensus saw the water- or gas-cooled reactors as just a temporary choice to be eventually replaced by liquid metal fast breeder reactors (LMFBRs). These reactors operate with fast neutrons produced by fuel that is enriched to a high degree with uranium-235 to convert the much more abundant but nonfissionable uranium-238, stored in a blanket surrounding the core, to fissile plutonium-239; the breeding should eventually produce at least 20 percent more fuel than it consumed. Because of its low cost and excellent heat-transfer properties, liquid sodium is the preferred coolant.

Leo Szilard had anticipated the breeder by 1943, and in 1945 Alvin Weinberg and Harry Soodak conceptualized the first sodium-cooled breeder design. A small experimental breeder coupled to a generator in Idaho Falls in 1951 produced the world's first fission-derived electricity; it first lit just four two-hundred-watt light bulbs and then the entire building in which it was located.[18] As the orders for American PWRs were reaching record numbers, Weinberg expressed confidence that "a nuclear breeder will be successful" and predicted it "rather likely that breeders will be man's ultimate energy source."[19] Westinghouse Electric also believed that nuclear power would offer "tremendous benefits in terms of greatly increased energy resources."[20] General Electric predicted in 1974 that the first commercial fast breeders would be operational by 1982, and that these reactors would account for half the new, large market for thermal generation in the United States by the year 2000.[21] GE forecast a rapid decline in the construction of new fission reactors (marked as light water reactors, LWR, in figure 2-3) after 1981, no new fossil-fueled generating capacity added after 1989, and the domination of America's electricity production by breeder reactors after 1992 (see figure 2-3).

Experts in other nations concurred with these expectations, and during the 1970s LMFBR projects were pursued by the Soviet Union, the United Kingdom, France, Germany, Italy, and Japan, as well as the United States. Completion of the American demonstration breeder was initially scheduled for 1975 at a cost of $100 million and was then postponed to 1982, with cost estimates reaching $675 million; the entire project was abandoned in

FIGURE 2-3

GENERAL ELECTRIC'S FORECAST FOR U.S. ELECTRICITY PRODUCTION

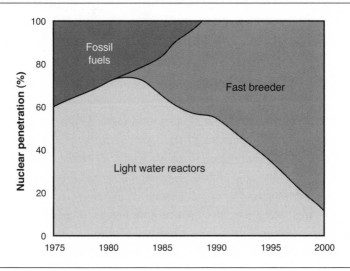

SOURCE: Based on Murphy (1974).

1983 as the declining cost of uranium and the rising cost of the reprocessing facilities used to separate plutonium from spent nuclear fuel made it clearly uneconomical.[22]

The French persevered and completed their full-scale breeder, the 1200 MW Superphénix at Creys-Malville, in 1986, but it operated at full power for less than ten months during the next eleven years. In February 1998, French prime minister Lionel Jospin confirmed its final shutdown. Japan's breeder was commissioned in 1994 but closed the very next year after more than 600 kg of liquid sodium leaked from its secondary coolant loop.[23] The history of technical advances does not offer many other examples of such a spectacular, multinational technical miscalculation. After decades of bold visions and careful and expensive preparation and development, and after tens of billions of dollars invested in prototypes, the breeder option was entirely stillborn. Research continues (now mostly concentrated in Asia, in China, India, Japan, and South Korea), but any truly commercial ventures are not on anybody's horizon.

In contrast, although the initially rushed adoption of conventional reactors stopped far short of early expectations of their market penetration, hundreds of them now work reliably in more than thirty countries, and nuclear fission has made a real difference globally, especially in some nations. In 2008, the world had 439 nuclear power plants with a total net installed capacity of about 371 GW—only about 11 percent of the global total. But nuclear reactors have load factors—that is, the percentage of time they are actually used to produce electricity—significantly higher than those of units powered by fossil fuels or water. Well-run nuclear power plants can operate 95 percent of the time, and the U.S. average is now nearly 92 percent, significantly up from about 75 percent in 1995.[24] This compares to typical rates of 65–75 percent for coal-fired stations, 40–60 percent for hydrogenation, and 25 percent for wind turbines. Because of this higher load factor, nuclear fission now generates about 16 percent of the world's electricity.[25] For some individual nations, the share is much higher—in France it is 78 percent, in Japan about 30 percent, and in the United States nearly 20 percent.[26]

Moreover, individual nuclear plants are among the largest electricity generation facilities. Installed capacity is commonly in excess of 1 GW, and in some cases as high as 5 GW, per plant, with individual turbogenerators rated mostly between 200 MW and 800 MW. In contrast, in 2007 the largest wind turbines were rated 5 MW, and the largest assemblies of photovoltaic cells added up to 4–6 MW of peak power. Nuclear fission is a reliable, high-capacity, high-load mode of electricity generation, which makes it an ideal complement to various renewable conversion modes that still have mostly low-capacity, moderate-load, and unpredictably intermittent operation.

New Case for Nuclear Energy

Nuclear power has gained many new converts because of its nearly carbon-free electricity generation.[27] Its new advocates include two well-known former adversaries, Patrick Moore, the cofounder of Greenpeace[28] and James Lovelock, the originator of the Gaia hypothesis (seeing the earth as a self-regulating superorganism).[29] They, and many others, believe that nuclear generation is the best and most readily available choice for preventing rapid and possibly catastrophic global warming. In fact, this is a highly questionable conclusion, as even tenfold expansion of the current nuclear capacity would avoid no more than about 15 percent of cumulative carbon emissions forecast to be released during the years 2000–2075.[30] Yet the case remains that the threat of global warming has changed many minds, and another round of oil price rises beginning in 2005 added a further incentive to go nuclear. Headlines, even in previously skeptical scientific periodicals, now indicate an increasing openness to nuclear power: "Is the Friendly Atom Poised for a Comeback?";[31] "Nuclear Reincarnation";[32] "Spinning a Nuclear Comeback."[33]

But has this reconsideration been enough to reverse the public's concerns about the safety of nuclear generation—another important obstacle standing in the way of its further development? Nuclear power's unique history, complexity, and dangers have meant that from its very origins it has been viewed hypercritically.[34] In the United States, public perceptions of unacceptable risk were boosted by the accident at the Three Mile Island plant in Pennsylvania in 1979 that damaged about 70 percent of a reactor core and melted about half of it.[35] Long-lasting European opposition to nuclear generation was hugely reinforced by an incomparably worse accidental core meltdown and the release of radioactivity amounting to about 5 percent of the reactor's radioactive core during the Chernobyl disaster in Ukraine in May 1986.[36] The fact that this accident arose from the combination of a flawed reactor design with no proper containment structure, unacceptable operating procedures, and inadequately trained personnel (a combination that did not obtain at any Western plant) became irrelevant as a radioactive plume drifted across the continent and contaminated large areas of Eastern and Northern Europe.

Nor has it mattered that actual health consequences were far less tragic than predicted.[37] The contrast between the repeated assurances of safety and the images of evacuees, sick children, and abandoned swaths of Ukraine was too powerful, and Chernobyl will continue to cast a long shadow over nuclear power for decades to come, particularly in Europe. Any serious student of comparative risk must feel exasperated that the years of analyses and operating experience proving Western nuclear power a highly acceptable choice apparently count for nothing. After 9/11, the fear of nuclear terrorism was added to perennial worries about the permanent storage of highly radioactive wastes, the real cost of nuclear generation, the chance of catastrophic reactor failures, near- and long-term environmental impacts, and the link between electricity from fission and nuclear proliferation.[38] Enrico Fermi recognized the first two concerns even before the end of World War II, when he suggested, "It is not certain that the public will accept an energy source that produces vast amounts of radioactivity as well as fissile material that might be used by terrorists."[39] Given that we have had more than half a century to act on these warnings, it is quite remarkable that more has not been done to address them.

If a new global nuclear era comes, it will have to be based on the better, more efficient, and inherently more reliable and safe designs that have been under development for more than twenty-five years.[40] Perhaps the most radical solution is for the reactors to be buried underground and to operate without any refueling, as was envisaged by Edward Teller in one of his last contributions to the field of nuclear physics.[41]

All this means that an early and substantial nuclear comeback is unlikely either in North America or in Europe. The only countries that continue to add significant nuclear capacity and plan further expansion are France (which based its successful nuclear program on a modular deployment of a Westinghouse PWR design), Japan, South Korea, India, Russia, and China. China has particularly bold plans, officially set at 40 GW of capacity by the year 2020 but recently estimated to reach perhaps as much as 70 GW in that year.[42] By comparison, the U.S. nuclear capacity is now just over 100 GW.[43] But much more is required if the world's nuclear industry is even to maintain its current share of electricity generation until 2030. Starting in 2009, a new 1 GW reactor will have to be built every sixteen days—a most unlikely prospect, particularly given the impact of the recent financial crisis on nuclear renaissance plans.[44]

Successful Failure

I have elsewhere called the use of nuclear fission for electricity generation a successful failure.[45] No other mode of primary electricity production was commercialized as rapidly as the first generation of fission reactors; only about twenty-five years elapsed between the first sustained chain reaction that took place at the University of Chicago on December 2, 1942, and the flood of new plant orders after 1965. But no other mode of electricity production has fallen so far short of its initial expectations.[46] And no other mode of energy production has received such generous public subsidies. U.S. data show that the nuclear industry was the recipient of no less than 96 percent of all funds, amounting to about $145 billion in 1998 dollars, that were appropriated by the U.S. Congress for energy-related research and development between 1947 and 1998.[47]

Nuclear electricity is now as important globally as hydroelectricity, and even relatively modest but steady capacity additions should keep that share, now close to 20 percent, from falling during the next ten to twenty years. But the economics of nuclear generation have always been in dispute, given the many externalities that have not been properly accounted for. They range from long-term health effects seen among uranium miners to the cost of decommissioning the shut-down reactors. Foremost among these concerns is the fact that no country, not even France with its bold commitment to nuclear generation, has thus far devised an acceptable method for permanently storing a relatively small volume of highly radioactive waste that must be sequestered for thousands of years.

This failure to date, of course, is not proof that effective solutions are impossible. It demonstrates only the enormous influence that a mistaken public risk perception can have on government policy, and also suggests the consistently inept bureaucratic handling of the challenge so far. This discouraging experience is even more incomprehensible given the fact that nuclear generation is the only low-carbon-footprint option that is readily available on a gigawatt-level scale. That is why nuclear power should be part of any serious attempt to reduce the rate of global warming; at the same time, it would be naïve to think that it could be (as some suggest) the single most effective component of this challenge during the next ten to thirty years. The best hope is for it to offer a modest contribution.

3

Soft-Energy Illusions

In October 1976, *Foreign Affairs* published a lengthy essay by Amory Lovins entitled "Energy Strategy: The Road Not Taken." The essay's epigraph was the last stanza of Robert Frost's poem "The Road Not Taken": "Two roads diverged in a wood, and I— / I took the one less traveled by." Lovins, too, chose the less traveled path and argued against the mainstream, business-as-usual strategy of U.S. energy policy, which stressed centralized conversions aimed at increasing the overall supply of energy, and particularly the generation of electricity. In contrast, the path Lovins chose—that of soft energy—combined "a prompt and serious commitment to efficient use of energy" with a "rapid development of renewable energy sources matched in scale and in energy quality to end-use needs."[1]

Lovins had no doubts about the eventual consequences of the choice. The large economic and sociopolitical problems associated with the first path would eventually become insuperable, while the second path, he argued, offered not just economic but also many social and geopolitical advantages, including the virtual elimination of nuclear proliferation around the globe. "It is important to recognize that the two paths are mutually exclusive," he wrote. "We must soon choose one or the other— before failure to stop nuclear proliferation has foreclosed both."[2] For Lovins, soft-energy technologies—a class that included photovoltaics as well as solar thermal devices— required no further technical elaboration— indeed, nothing extraordinary or, as he put it, "exotic." He concluded that a largely or entirely solar economy could be constructed in the United States with available soft-energy conversions that were either already, or nearly, economical.

Advantages of Soft Energy

Lovins made many general claims with a confidence that allowed no room for doubt or qualification. Most notably, he maintained that "renewable energy flows . . . are always there whether we use them or not," and he argued that soft-energy conversions "are matched in scale and in geographic distribution to end-use needs, taking advantage of the free distribution of most natural energy flows."[3] Decentralized electricity generation, he asserted, could reduce—even eliminate—fixed distribution costs (ranging from transmission lines and transformers to billing computers and office workers), and these savings would far outweigh the costs of maintaining the dispersed infrastructure of small systems. But smallness was always inviolate—when Lovins admitted that not all energy systems need be at domestic scale, he allowed only for medium-scale systems serving neighborhoods or villages, not for anything strongly centralized.

He also foresaw, again quite confidently, many specific advances, including "exciting developments" in converting agricultural, forestry, and urban wastes to methanol and other biofuels, which would be "sufficient to run an efficient U.S. transport sector."[4] For Lovins, the value of the soft-energy path reached far beyond its powerful yet benign solutions to the world's energy problems. Harnessing these energies would be not only economical but also "elegant," a quality cherished by engineers; and small-scale, decentralized conversions would particularly benefit the poor, since they would contribute "promptly and dramatically to world equity and order." Indeed, such conversions would foster the diffusion of democracy "from the ground up," even as they spread the virtues of community resilience and self-sufficiency and provided safe and "ecologically inoffensive" alternatives to an inherently destructive and risky hard-energy path.[5]

Although the paper had thirty-six citations, it did not refer, most curiously, to what was clearly one of its major inspirations. Ernst F. Schumacher's *Small Is Beautiful* was a slim volume by a British economist, statistician, and long-time adviser to the National Coal Board that rapidly established its author as the most influential advocate of smallness.[6] His approach to economic development was based on four fundamental dicta: Make things small where possible, reduce capital intensity, make the process simple, and make it nonviolent.

Schumacher's work was a key theoretical justification for a new form of a globally applicable economic development strategy whose critical ingredient was what came to be known as intermediate or appropriate technology. The invariably small-scale and simple techniques and methods associated with this technology stood in obvious contrast to large-scale, high-tech approaches, which were seen as grossly mismatched with the enormous needs (whether for jobs, food, or energy) of poor, populous countries. Without any doubt, Lovins's *Foreign Affairs* article should be seen as a specific application to energy affairs of the general Schumacherian principle.

But before addressing the substance of the paper, I must first point out that Lovins misrepresented Frost's poem. His interpretation implies a stark duality between the two opposite energy paths, one hard and one soft, but the poet wishes he could have it both ways ("And sorry I could not travel both / And be one traveler, long I stood); that is why he felt that he was not deciding between stark opposites ("Then took the other, as just as fair"), but rather making a choice that was difficult.

The only key point that Lovins got right in his essay is that the version of the hard path he portrayed did not become a reality. Of course, forecasting this was no great achievement; it was merely a matter of outlining an extreme prospect (that is, the most extravagant version of the myth of nuclear power) to argue that it should not, and almost certainly would not, happen.

Lovins's career started during the early 1970s with an aggressive opposition to the nuclear industry, a vigorous dismantling of exaggerated expectations for its growth, and a detailed critique of the safety, environmental, and political implications of relying primarily on nuclear fission. This background enabled Lovins to expose the extreme hard-path myth, including a simplistic long-term energy forecast of unchecked continuation of previous exponential growth that implied the existence of as many as 800 nuclear reactors by the year 2000, when the United States would consume more than 150 EJ of primary energy. Fifteen years after his *Foreign Affairs* essay, Lovins looked back and confirmed that the "hard path hasn't happened and won't."[7] True, at that time the world's coastlines were not dotted with nuplexes housing multigigawatt reactors; the global count of fission reactors had not reached many thousands; and fast breeders were not the leading source of new electricity.[8] In that regard, Lovins was absolutely right.

But he did not call attention to an egregious forecasting failure of his previous article—namely, the fact that the trajectory of the soft path he had proposed was even more remote than it had been in 1976. In 1976, he believed that renewable conversions would "quickly emerge to displace much of the oil and gas we currently consume," but in 1992 there was no sign of such an epochal transition, so he was left only to wonder about the emergence of renewable energy conversions that would displace the hydrocarbon consumption. And nearly two decades later there has been no fundamental change in this situation.

Soft Energy Today

At the beginning of the twenty-first century, no major economy relied on soft (small-scale, decentralized, renewable) energy conversions for anything more than a negligible fraction of its primary energy supply. Lovins anticipated that an alternative soft-energy future would require only about 100 EJ in the year 2000, a total virtually identical with the actual U.S. demand in that year and a third lower than the unrealistically posed hard-path scenario. While the estimate for aggregate performance was fairly close, however, the details were wrong: Lovins failed, and rather spectacularly, when he predicted the likely composition of this energy supply. His breakdown was about 29 percent of the total coming from coal, 33 percent from oil and gas, and 33 percent from soft-energy conversions. In reality, about 23 percent came from coal, 62 percent from oil and gas, 8 percent from nuclear fission, and 3.2 percent from conventional hydrogeneration.[9]

Renewable energies other than large-scale hydroelectric power (an energy conversion that, according to Lovins's own criteria, obviously does not belong to the soft path) provided just over 4 percent of all U.S. primary energy consumption. In fact, more than 90 percent of this total was accounted for by the burning of logging residues in wood-processing and pulp and paper enterprises, by ethanol production in large-scale industrial facilities, and by generation of electricity in large commercial wind farms. In other words, small-scale, decentralized energy conversions contributed less than 0.5 percent of the U.S. primary energy supply in 2000, rather than the 33 percent envisaged by Lovins in 1976, or less than 0.5 EJ rather than 33 MJ.

Missing a target by 98.5 percent—that is, getting it wrong over the course of twenty-four years by a factor of more than sixty—does not constitute brilliant foresight. The indisputable disappearance of the original soft-energy mirage is the most telling illustration of the vastly unrealistic hopes that were invested in the less-trodden energy future during the late 1970s.

The Hypercar

The soft-energy vision receded during the 1980s, and by the early 1990s Lovins was back with another bold claim, the rapid and all-conquering emergence of the Hypercar. That aerodynamically superslippery machine was to be built like an airplane from superlight carbon fiber stronger than steel; was to be quiet, safe, and 95 percent less polluting than conventional vehicles; was to be cheap to lease or rent; and, best of all, was to have a fuel efficiency equal to as much as 200 mpg.[10]

The Hypercar Center was set up in 1994 to research and promote the proposed car, and a for-profit venture, Hypercar Inc., was launched in 1999. In December 2000, when asked how the Hypercar was doing, Thomas Crumm, the CEO and president of the company, said, "It's doing fine. We are on schedule, we're on budget." He promised that a site for a new factory would be revealed within eighteen months.[11] But no site was revealed, and there is no factory or 200 mpg Hypercar; in 2004 the company changed its name to Fiberforge "to better reflect the company's new direction and its goal of lowering the cost of high-volume advanced-composite structures."[12] So much for the car that would be cheap, clean, and efficient.

Other Soft-Energy Dreams

Lovins was not the only enthusiastic promoter of a soft, decentralized energy nirvana. A massive report of nearly 1,800 pages by the InterTechnology Corporation concluded that solar energy could supply 36 percent of America's industrial process heat by 2000, including at least 70 percent of all heat for applications requiring temperatures less than about 300°C.[13] A Harvard Business School study suggested that by the year 2000 the United

States could "reasonably" satisfy 20 percent of its total energy needs through solar sources, with solar heating, both active and passive, being the single largest contributor; no new conversion techniques would be needed to achieve that goal.[14] In the same year, Hayes was a bit more conservative, envisaging about 25 percent of all U.S. energy coming from decentralized renewable resources within the next fifty years.[15]

In contrast, Sørensen (1980) predicted that 49 percent, or almost half, of America's energy would come from renewables by the year 2005. He specified that wind and biogas—a gas produced by bacterial breakdown (anaerobic digestion) of organic matter—would each supply 5 percent of the total, and that decentralized photovoltaic electricity would add 11 percent. As it happened, the biogas share was less than 0.001 percent of America's primary energy in 2005, wind turbines produced less than 0.2 percent, and photovoltaics added less than 0.01 percent.[16] In other words, Sørensen's forecasts erred by three orders of magnitude.

Europeans had their own soft-energy dreams. For example, Johansson and Steen were confident that by the year 2015 Sweden would be wholly capable of managing its energy supply using renewable domestic energy sources.[17]

Soft Energy in China

Schumacherian smallness in general and Lovinsian energy softness in particular found unexpected support in Maoist China in the 1970s and the early 1980s.[18] Mao's thought in action preceded Schumacherian preaching by fifteen years: The Great Leap Forward, launched in 1958, was based on a delusionary idea that a poor, underdeveloped nation could catch up with the world's most advanced economies in a single, frenzied spurt of a few years.

This impossible goal involved mass replication of primitive, small-scale techniques, with hundreds of millions of people forced to cut down trees, mine poor iron ore and coal, and build primitive backyard furnaces to smelt substandard iron. But this leap ended in the worst man-made famine in history, in which more than 30 million people died, and in a reversion to normal economic practices.[19] The excesses of the late 1950s were not repeated during the 1970s, but smallness and simplicity once again influenced energy policy in China. Small coal mines, small hydrostations, and family-sized biogas digesters were constructed and operated during this period. The latter two were perfect embodiments of renewable energy softness, and, as such, they garnered plenty of uncritical admiration by the Western devotees of smallness.

The practice of small-scale biogas generation was based on eclectic inputs available even in poor villages. Animal dung, human feces, pieces of vegetation (crop stalks, straw, grass clippings, leaves), garbage, and wastewater were sealed up in insulated brick or concrete containers (digesters) and left to decompose. Biogas produced by anaerobic methanogenic bacteria is 55–70 percent CH_4 (methane) and 30–45 percent CO_2, and its energy content is 22–26 MJ/m^3. Villagers used it for cooking and lighting, and a typical 10 m^3 digester was claimed to cover these needs for a typical south Chinese family of five; promotion of the practice began in the early 1970s in Sichuan, where more than 30,000 digesters were built by the end of 1973 and more than 400,000 of them were reported to be in operation by the middle of 1975.[20]

In 1978, China's official goal was to have 20 million digesters in 1980 and 70 million units by 1985. But once their diffusion ceased to be a subject of Maoist campaigns—and once the peasants, freed by Deng's reforms to manage their activities for profit, saw them from a purely economic perspective—the bubble burst. As rural China sought to improve its standard of living, and as villagers began to engage in various private activities, the digester total fell below 4 million by 1984. Although the total later rose a bit because many farmers found the use of larger digesters for animal waste control profitable, it never surpassed the 1979 peak. Moreover, most of the remaining biogas digesters were unable to produce enough fuel to cook rice three times a day, still less every day for four seasons.[21]

The reasons were obvious to anyone familiar with the complexities of bacterial processes. Biogas generation, simple in principle, is a fairly demanding process to manage in practice. The slightest leakage will destroy the anaerobic condition required by methanogenic bacteria; low temperatures (below 20°C), improper feedstock addition, poor mixing practices, and shortages of appropriate substrates will result in low (or no) fermentation rates, undesirable carbon-to-nitrogen ratios and pH, and the formation of heavy scum. Unless it is assiduously managed, a biogas digester can rapidly turn into an expensive waste pit, which—unless emptied and properly restarted—will have to be abandoned, as millions were in China. Even widespread fermentation would have provided no more than 10 percent of rural household energy use during the early 1980s, and once the privatization of farming got underway, most of the small family digesters were abandoned. By 1990 China had hardly any digesters, but it had become essentially self-sufficient in food—moreover, at a daily per-capita rate almost as high as Japan's.[22]

Small-scale hydrostations made more sense. Thanks to its mountainous terrain, no other country has a higher hydrogenerating potential than China, and by building small stations the Chinese were only following many historical precedents, as such installations were common during the early stages of electrification in North America, Europe, and Japan. China's small hydrostation program began as part of a massive water conservancy effort during the Great Leap years, and Maoist planners had visions of no less than 2.5 GW of aggregate capacity in 1967. In reality, when the Great Leap collapsed in famine, the total amounted to less than 500 MW.[23]

A new wave of construction began during the 1970s when, once again, traditional methods of mass labor construction were used to build small rock-filled or earth-filled dams requiring only a minimum of cement, steel, and timber. Their numbers rose from 26,000 in 1970, when the mean size was just 35 kW, to about 90,000 by 1979, when it was 70 kW. After this, their total fell by 20 percent in five years, while the size of the remaining stations increased significantly. Projects of several MW became relatively common as China's newfound appreciation of costs and rational economic management sidelined the worship of smallness for its own sake.

This retreat from smallness was due partly to the types of problems that beset small energy projects in general, and partly to specific Chinese and environmental factors. Many hastily built stations were simply shoddy structures that leaked or collapsed. Repeated drought caused complete desiccation of many tiny reservoirs, while accelerated silting destroyed the small storage capacities of others. Even in normally wet years, the average load factor of these stations was only about 25 percent, or about 2,200 hours a year, compared to 4,000–4,500 hours for larger plants. And the capital cost per unit of capacity was often significantly higher than in larger projects.

Less than a decade after the end of Maoism, the Chinese pendulum had swung dramatically the other way, and that trend has only intensified in recent years. Gone are the campaigns promoting tiny biogas digesters and small hydrostations of less than 50 kW. The country that was once seen as the greatest potential beneficiary of small-scale soft-energy conversions and the most convincing embodiment of the Schumacherian future has become a serial builder of energy megaprojects. In 2006, China commissioned the equivalent of ninety large (1 GW) coal-fired electricity generating plants, or nearly as much as the entire French capacity. It also completed the world's largest hydrostation, Sanxia on the Yangzi, with 18.2 GW—about 45 percent larger than the second-largest project, Itaipú on the Paraná between Brazil and Paraguay. By any criteria, the hard-energy path has fared only too well in China.

The "Perfect" Solution

My point is not that there is anything intrinsically wrong with smallness and softness. There is no single "correct" scale of capacity or complexity for energy conversions, and diversity and variety are important elements in any energy policy. It is the ideological worship of scale that is wrong. I have always thought it simply counterproductive to exaggerate any future contribution of a decentralized energy supply, be it in affluent societies or in modernizing economies. Depending on circumstances, small and soft may indeed be beautiful and desirable—but, clearly, nothing is inherently superior in that approach. In the real world, there are inherent and predictable, as well as hidden and surprising, advantages and drawbacks to scales small *and* large; judging a technique solely by its scale is neither rational nor useful.

What has led me to view the promise of small, decentralized renewable energy conversions so skeptically, and to judge its leading promoters so harshly, is not their enthusiasm for a "perfect" solution (unfortunately fairly common), but all those unrealistic assumptions, patent exaggerations, and irresponsible claims made on behalf of these resources and techniques. Lovins certainly knew that renewable resources are not "always there whether we use them or not."[24] The sun-drenched tropics, for example, turn out to be not that sunny. The fact is that most of coastal Nigeria and the Brazilian Amazon receives less radiation annually than Georgia or Kansas, and nearly all densely populated regions of Southeast Asia, from

China's southernmost provinces to Sumatra and Kalimantan, have annual insolation comparable to that of northern France and southern England—locales that are not usually perceived as "sun-drenched."

Lovins's statement that renewable energy flows are "matched in scale and in geographic distribution to end-use needs" is similarly misleading, as the Chinese found after wasting so much effort on an ideologically enforced soft-and-small approach. More than half of humanity is now living in cities, and an increasing share inhabits megacities from São Paulo to Bangkok, from Cairo to Chongqing, and megalopolises, or conglomerates of megacities.[25] How can these combinations of high population, transportation, and industrial density be powered by small-scale, decentralized, soft-energy conversions? How can the fuel for vehicles moving along eight- or twelve-lane highways be derived from crops grown locally? How can the renters of smallish cubicles on the thirtieth floor of high-rises—facing even taller walls just a few meters away—extend their individual solar heaters or wind turbines from their windows? How can the massive factories producing microchips or electronic gadgets for the entire planet be energized by attached biogas digesters or by tree-derived methanol? And while some small-scale renewable conversions can be truly helpful to a poor rural household or to a small village, they cannot support such basic, modern, energy-efficient industries as iron and steel making, nitrogen fertilizer synthesis by the Haber-Bosch process, and cement production.

In 1978, Lovins claimed that "soft technologies are . . . increasingly used in practice, to construct smooth transitions (over 50 years or so) to virtually complete reliance on appropriate renewable energy sources."[26] As I have quantified in detail, three decades into this transition (excluding large-scale hydrogeneration, which has been a well-established, centralized, hard technique for decades), the United States derives less than 1 percent of its primary energy from new renewables and less than 0.1 percent from smaller, decentralized conversions. Centralized electricity generation still dominates, and there are no signs of its imminent retreat. And all those relatively small contributions by renewables are based on larger, and increasingly numerous, commercial installations. The current biofuels craze relies on the large-scale industrial conversion of cane and corn, not on any household-size units. Wind installations are now packing ever larger individual turbines into projects rated at hundreds of megawatts, onshore and offshore.

The Future of Soft and Small Approaches

The overall energy supply draws a bit more on renewable flows, but hardly on the small, decentralized units of the soft vision. The verdict is clear: Soft and small has not worked as predicted. In the United States, soft and small will not expand from its tiny base to fill the country's entire demand in the less than two decades remaining on Lovins's initial transition schedule, according to which soft conversions were to cover the total U.S. energy demand by 2025. And in the populous, rapidly modernizing, low-income countries (now by far the fastest growing segment of global energy use), soft and small does not, contrary to the original claims, fit the scale, which is increasingly mega-urban with high densities of steadily rising round-the-clock demand.

What is perhaps most curious, and most counterintuitive, about the soft-energy vision is how much it had in common with the hardest of all conventional energy paths, and the one it was designed to eliminate—that is, the all-encompassing vision of the nuclear future. They shared a misplaced faith in technical fixes as the best solution to the complex challenge of ensuring a global energy supply. Ignoring inconvenient realities, they made claims for those techniques based on wishful thinking, proceeding without any solid demonstration that such techniques, whether fast breeders or biofuels, could be developed rapidly for widespread commercial use at an affordable cost.

The failure of the soft-energy vision is not surprising, once one understands that its genesis, its original appeal, and its uncritical welcome were products of the countercultural revolt of the late 1960s and the early 1970s. This previously untrodden path was to effect a profound social transformation, not just a new way of securing energy. Lovins said so explicitly: "Perhaps the most profound difference between the soft and hard paths is their domestic sociopolitical impact."[27] And so we can dispose of the small-scale, decentralized soft-energy vision in a historically appropriate way: Add it to the list of modern grand schemes intended to reform society and demonstrate the brilliance of theoretical ideas—from American Technocracy, popular during the 1930s and advocating individual energy consumption coupons, to the Maoist path to modernization that derailed China's development for three decades. Their common denominator: failures all.

PART II

Myths in the Headlines

The three myths I considered in part I are old but enduring. In part II, I look at five more recent myths—ones that are making headlines right now.

The context for these newer myths is the dichotomous view of energy sources now so widely held: Fossil fuels are bad, renewable energies are good. While fossil fuels remain the very foundation of modern economic growth, spreading prosperity and a decent quality of life, they are no longer seen in that light. Rather, they are perceived as undesirable, outright dangerous, or even immoral, since their continuing use is thought to pose an unprecedented threat to the survival of modern civilization. Growing fears about rapid global warming caused by emissions of CO_2 from the combustion of fossil fuels are behind this increasingly stringent judgment, and these fears feed (mostly unrealistic) visions of an accelerated global transition to nonfossil energies.

Coal has always been more polluting in terms of particulate matter and sulfur oxide emissions than other hydrocarbons, and because it also has the highest CO_2 emissions per unit of released energy, it is seen as the most undesirable choice. A closer look at coal's attributes and the history of its use shows that this judgment is unfair and suggests that if the fuel's conversion were done with the most efficient techniques available today, we would have no reason to view it so negatively. Crude oil—largely because of the continuing indispensability of refined fuels for the entire transportation sector—occupies a more exalted place than coal. Although its considerable environmental impact is a concern, the main worry about oil is that its global extraction may peak in the very near future, and that this peak will not be followed by a prolonged production plateau but, rather, by a steep decline that will bring a multitude of economic and social hardships—in the most extreme versions, the end of modern civilization.

That is why the first myth I debunk in this part of the book is the peak oil myth. I review the arguments for the imminence of this epochal event, as well as arguments about how to postpone it or lessen its impacts. My verdict does not support a fashionable notion of early, inevitable, and pronounced extraction declines. I show instead that the share of conventional oil in the global energy supply will gradually decrease, but that hydrocarbons, liquid and gaseous, will remain a major source of energy for decades to come.

This development will be mainly due to the combination of a rising importance of natural gas and increasing recovery of nonconventional oil resources. Except for flying, everything that we now do with liquid fuels—in transportation, heating, and industrial production—can be done with natural gas, and gradual substitutions of gas for liquid fuels would not require any untried or unaffordable adjustments. And this substitution would be based on what are still substantially growing natural gas reserves: despite an unprecedented increase in consumption, they rose by nearly 70 percent between 1988 and 2008 and are already nearly as large (90 percent in 2008) as the reserves of crude oil.[1] Moreover, their total is expected to continue rising at an impressive rate, as technical advances (horizontal drilling, hydro-fracturing of gas-bearing rocks, 3-D seismic imaging) have opened up new reservoirs of this clean and versatile fuel and created, for the first time in history, a truly global natural gas market.

As for the nonconventional oils, they are found as barely liquid matter in heavy oil in many basins around the world, including Canada and the Middle East; as solid hydrocarbons in enormous tar deposits, particularly those of the Orinoco Belt in Venezuela; and as solids dispersed in relatively low concentrations of 5–20 percent among sedimentary rocks.

Oil recovery from Alberta's oil sands began on a small scale during the 1960s, and by the beginning of the twenty-first century, the combination of rising crude oil prices and widespread anticipation of declining production of conventional oil made the development of this nonconventional oil resource one of the most sought-after investment opportunities in the oil industry. Realistic assessments show Alberta's oil sands to be an energy resource of commercial interest—but not, as some enthusiasts claim, any replacement for Saudi Arabian oil. Thus a myth has arisen within a myth, and below I deconstruct both. The oil era will not end very soon or very precipitously, and oil extracted from Alberta's sands will not make up for the oil currently extracted from Saudi Arabia's giant and supergiant oilfields.

The second myth I examine in part II is not a vision of catastrophe, like the myth of global peak oil extraction, but rather a vision of salvation: the potential for a long-lasting continuation of fossil-fueled civilization through carbon sequestration. Now considered a serious, efficacious method for extending the age of fossil fuels while reducing—eventually even eliminating—additional CO_2 emissions in the atmosphere, carbon sequestration supposedly would turn fossil fuels once again into acceptable, indeed innocuous, sources of energy. I show that the world of carbon-free combustion is not around the corner. I also demonstrate why it would be very difficult to capture and hide enough CO_2 during the next two to three decades to slow dramatically the increase of atmospheric concentrations of this greenhouse gas, or even stabilize them at a level compatible with only a minor increase of average tropospheric temperature.

The next two myths I debunk have to do with the perception of renewable energy flows as a perfect "green" salvation, generating no carbon, sparing the environment, stimulating new economic activities, and providing the best sustainable energy foundation for modern civilization. Indeed, few energy topics are subject to as many interpretations and persistent misconceptions as renewable energy resources and their conversions.

The difference between resources (the total presence of a natural commodity) and reserves (the small part of that total that is economically recoverable at any given time) is not always well understood or appreciated; but it is obvious that these two fundamentally different categories are chronically conflated in the minds of renewable energy enthusiasts. Or, perhaps more accurately, the enthusiasts are simply unaware that such a distinction applies no less to wind or straw than to crude oil or oil sands. Hence, they go on quoting the highest available resource estimates as if they were the flows readily suited for commercial exploitation. We read, for example, that just 1 percent of the jet stream's wind power could supply all the world's energy needs—but we are told nothing about the minor matter of how to convert affordably winds of more than 100 km/h, and often faster than 200 km/h, blowing 10–12 km aloft at the altitude of intercontinental jet flights, into any useful form of commercial energy.

I focus on just two fashionably misrepresented and widely misunderstood renewable conversions: liquid fuels made from biomass (trees, crops, and their residues) and wind-powered electricity generation. These choices are easily justified by the intensity of recent interest—by governments seeking to lower their countries' dependence on imported fuels; by industries looking for lucrative new opportunities; by citizens worried about an imminent peak of global oil production; and, of course, by all shades of green activists who see such energy conversions as near-miraculous goods that do not run out or generate greenhouse gases or, indeed, any form of pollution, even as they take away a dangerous oil weapon from untrustworthy Middle Eastern regimes and foster self-reliance and decentralized energy production.

Clearly, there is nothing modest about the many claims made for these renewable conversions. Green energy enthusiasts do not envisage them as making only an important contribution, supportive but not decisive—providing, say, 15–20 percent of national, regional, or global supply. Instead, they confidently propose a new world where biomass will be the leading source of primary energy supplies, perhaps in just two generations; where biofuels produced by the fermentation of carbohydrates, directly or after hydrolyzing cellulose into its constituent glucose molecules, will entirely supplant liquid fuels refined from crude oil; and where wind, whose theoretical global energy potential is a multiple of the world's existing demand, will be used to generate most electricity.

The genesis and acceptance of these myths spring from the same combination of deficits: a naïve understanding of the underlying biology, chemistry, and climatology; simplistic and indefensibly optimistic assumptions regarding average potentials and performances; and ignorance of the length of time required to effect any fundamental energy transition.

At this point, I should make an important prefatory declaration: I am very much in favor of all sorts of renewable energy conversions. The most general justification of this position rests on a well-proven ecosystemic principle that is equally applicable to human societies: A more diverse supply fosters a more stable system. Other broadly applicable justifications include comprehensive and less wasteful uses of already exploited resources and the advantages of harnessing bountiful local, regional, or national energy flows for specific renewable conversions.

These conversions, however, must be viewed through the same prism as are fossil fuels. I see major resource constraints, complications in harnessing the flows, questionable economics (fossil fuel accounting does not have a monopoly on ignoring externalities), and undesirable and unintended side effects and consequences, some immediate, others apparent only after longer periods of time. All these constraints suggest that supporters of such conversions are irresponsibly exaggerating their promise, and that the belief in these conversions as imminent agents of a great epochal deliverance—the path to a civilization sustained on a new energy basis—is, at best, naïve. They should become an important part of an energy solution, but they cannot make as much difference, and as fast, as is now so commonly believed.

The last myth I address in some detail concerns the pace of energy transitions—the amount of time needed to shift from the prevailing composition of primary energy sources, be it on a national or global level, to a new mixture of primary energies, or from dominant energy prime movers (be they internal combustion engines in car transportation or steam turbo-generators in centralized electricity generation) to new conversion techniques. Historical evidence shows that energy systems, the most complex and capital-intensive mass-scale infrastructures of modern societies, are inherently highly inertial, and that our determination can accelerate their change but cannot fundamentally alter the gradual nature of their evolution. This means that such goals as repowering America in a decade are nothing but frivolous suggestions with no chance of actual realization. What is needed is a realistic quest combining reduced energy use, improved conversion efficiencies, and gradual introduction of new ways of harnessing and using all kinds of non-fossil energies.

4

Running Out: Peak Oil and Its Meaning

"I am a school librarian who has been researching peak-oil theory with my students at our school," the e-mail began, "and we have been very disturbed by the forecasts of peak-oil theorists, especially by the doomsday scenarios put forth by them about mass unemployment, starvation, and the near-stone age conditions they predict in the coming decades. . . . I fear for my own children, as well as for the students I have been working with on this topic." Fortuitously, I got this message from a librarian in upstate New York after she had read one of my papers on peak oil during the same week I began to write this book, and her reaction is understandable, given the content and tone of publications and speeches predicting an imminent peak in global oil extraction.

Recent concerns about an imminent end of the oil era, formulated as the theory of peak oil, began with a group of retired geologists, including Colin Campbell, Jean Laherrère, L. F. Ivanhoe, Richard Duncan, and Kenneth Deffeyes. The movement began during the 1990s and during the first decade of the twenty-first century writings and speeches of these individuals created a mass following that has gathered under the www umbrellas of peakoil.net, peakoil.com, peakoil.org, and hubbertpeak.com, and, more professionally, in the Association for the Study of Peak Oil (ASPO). Peter Odell, an astute lifelong observer of the global oil scene, calls its adherents peak-oilers, and they have been flooding the media with catastrophist tales describing the consequences of a precipitous decline in oil availability. According to Ivanhoe, for example, the end of the oil era in the near future will bring "the inevitable doomsday" to be followed by "economic implosion," which will make "many of the world's developed societies look more like today's Russia than the U.S."[1]

But nobody has projected the consequences as far as Richard C. Duncan in his Olduvai Gorge theory. Duncan sees a global oil production peak as nothing less than "a turning point in human history"—the beginning of a

FIGURE 4-1a

THE OLDUVAI GORGE THEORY: DUNCAN'S DEPICTION OF THE OIL PEAK

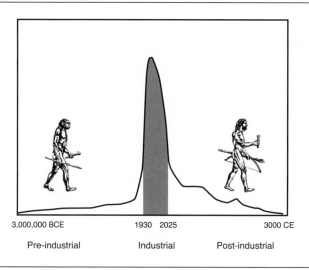

SOURCE: Based on Duncan (1998, figure 1).

FIGURE 4-1b

AVERAGE GLOBAL PER-CAPITA CONSUMPTION OF PRIMARY ENERGY, 1950–2009

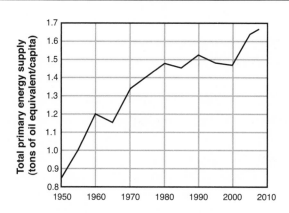

SOURCE: Averages were calculated using energy data in British Petroleum (BP, 2009) and population totals in United Nations (2008).

rapid decline in oil production that will effect the demise of industrial civilization, beginning as early as 2025. The results will be massive unemployment, breadlines, homelessness, and the end of industrial civilization, ultimately returning humanity to a life comparable to that experienced by some of the first primitive hominids inhabiting the Kenyan gorge some 2.5 million years ago.[2] This theory is built on Duncan's mistaken claim that the average per-capita global consumption of energy peaked in 1978 and has been declining ever since (see figure 4-1a).[3] In reality, in 2008 the average global per-capita consumption of primary energy was nearly 10 percent higher than in 1978 (see figure 4-1b),[4] but even a lower rate would not signal anything catastrophic; because of the steadily falling energy intensity—the energy consumption per unit of economic product—of the global economy, it could actually be a sign of progress for the world to use less energy.

Predictions of Peak Oil Production

Although this deep catastrophism, utterly unjustified by objective evidence, is fairly new in oil resource debates, the concern about an imminent exhaustion of mineral resources in general, and of crude oil in particular, is not. The earliest published concerns about an approaching end of oil production go back to the 1870s; they were voiced again by the director of the U.S. Geological Survey during the early 1920s, less than a decade before the discovery of the supergiant West Texas oilfield in 1931. The most famous post–World War I oil peak forecasts are those of M. King Hubbert from the late 1950s and 1960s. Hubbert assumed that the extraction of any mineral resource follows an exhaustion curve that conforms to the normal symmetrical, bell-shaped distribution; there is no prolonged plateau, and the peak output is rapidly followed by a decline whose course mirrors the production rise.

Hubbert is the patron saint of peak-oilers, considered an infallible and prescient seer because, as we are repeatedly reminded, he correctly predicted the peak of U.S. oil production in 1970. A closer look shows a much less impressive record. In his March 8, 1956, presentation before the spring meeting of the Southern District Division of Production of the American Petroleum Institute, Hubbert plotted two production curves for the United

FIGURE 4-2a

**HUBBERT'S PREDICTIONS FOR U.S. CRUDE OIL PRODUCTION
VS. ACTUAL PRODUCTION**

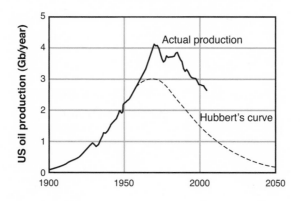

SOURCE: Based on Hubbert (1969) and on extraction data in U.S. Department of Energy, Energy Information Administration (2009).

States, one based on an ultimate output of 150 billion barrels, which would peak in 1962 at 2.6 billion barrels a year, and another for an ultimate output of 200 billion barrels, which peaked in 1968 at 3 billion barrels a year.[5] In later revisions of this original work, he put the peak of the complete cycle of U.S. petroleum liquids (that is, crude oil and natural gas liquids) at "about 3.5 billion barrels a year . . . during the first half of the 1970-decade."[6]

The actual peak did come in 1970, but at 4.12 billion barrels—18 percent above Hubbert's prediction. More important, Hubbert's peak estimate was, as indicated, based on the ultimate recovery of 200 billion barrels total, but between 1859 and 2005 the U.S. oil industry had already produced 192 billion barrels. The industry is still the world's third-largest producer of crude oil, and at the end of 2008 it still had 30 billion barrels of remaining reserves. Thus, the post-peak decline of U.S. oil extraction has not been a mirror image of the incline: Hubbert's rate for the year 2000 production was 1.5 billion barrels, while the actual extraction was 2.8 billion barrels, or nearly 90 percent higher—hardly an enviable accuracy for a thirty-year forecast. Actual output in the year 2008 was about 75 percent above the rate forecast by Hubbert[7] (see figure 4-2a).

FIGURE 4-2b

HUBBERT'S PREDICTIONS FOR GLOBAL CRUDE OIL PRODUCTION VS.
ACTUAL PRODUCTION

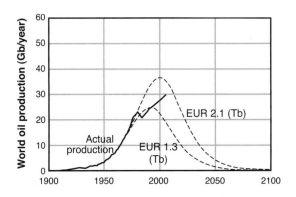

SOURCE: Based on Hubbert (1969) and on extraction data from British Petroleum (BP, various years).
NOTE: "EUR" signifies values for "estimated ultimately recoverable" oil.

Hubbert's failures have been even more glaring as far as the forecast of global peak oil extraction is concerned. In 1969, he offered two different estimates of ultimately recoverable oil: either at 25 billion barrels in 1990 or at 37 billion barrels in 2000. Both estimates projected a symmetrical curve and continuing high demand such as prevailed during the 1960s. He could not have anticipated the substantial decline in oil demand following OPEC's two rounds (1973–74, 1979–81) of extortionary price increases. Global oil extraction did not peak either in 1990, when it was actually about 4 percent below the level forecast by Hubbert, or in 2000, when it was at 27.4 billion barrels, 26 percent lower than Hubbert's predicted peak. By 2008, it was still just below 30 billion barrels[8] (see figure 4-2b). In this case, Hubbert was nowhere near being correct either on the timing or on the production level.

This is not surprising, because the symmetrical model of oil extraction is just one of many possibilities, and we now have a rigorous quantitative proof that it is not either a dominant or a modal choice. Brandt tested the assumptions of the Hubbert method by analyzing oil recovery data from 139 spatial units, ranging from the state and regional level in the United States to subcontinental and continental scales. He concluded that while no model (symmetrical, asymmetrical, linear, exponential) dominates, "When attempting to understand past production, symmetric models are not satisfactory," and once asymmetry is allowed in the production curves, then "the asymmetrical exponential model becomes the most useful model."[9]

Hubbert's followers have not fared much better with their forecasts. In 1977, the Workshop on Alternative Energy Strategies predicted the global oil peak as early as 1990 and most likely between 1994 and 1997.[10] In 1978, Andrew Flower wrote in *Scientific American* that "the supply of oil will fail to meet increasing demand before the year 2000."[11] A year later, the Central Intelligence Agency (CIA) concluded that the global output must fall within a decade.[12] Some of the latest peak oil predictions have already failed: Colin Campbell's first global peak was to be in 1989, Ivanhoe's in 2000.

Certainly the most bizarre prediction was offered by Kenneth Deffeyes, an experienced petroleum geologist and a former professor at Princeton University. As a scientist, Deffeyes must know that the real world is permeated by uncertainties, that complex realities should not be reduced to simplistic slogans aimed at gaining media attention, and that, as even a brief retrospective will demonstrate, making precise predictions is a futile endeavor. Nonetheless, he treats peak global oil production in a way that leaves no room for any doubt ("No initiative put in place starting today can have a substantial effect on the peak production year"), that portrays the world's energy use as merely a matter of supply and utterly ignores demand, and that goes farther than any of his confrères by predicting not just the year (although originally he said it will come in 2003) but also the very day when the world's oil output is to peak. Admitting that doing so was "a bit silly," he nonetheless asserted in 2004 that "we can now pick a day to celebrate passing the top of the mathematically smooth Hubbert curve: Nov. 24, 2005. It falls right smack dab on top of Thanksgiving Day 2005."[13]

The fundamental problem with the notion of predicting a peak for oil extraction is that it rests on three simple assumptions—that recoverable oil resources are known with a high level of confidence, that they are fixed, and that their recovery is subsumed by a symmetrical production curve—which happen not to be true. These three claims mix incontestable facts and sensible arguments with indefensible assumptions, and they caricature complex processes and ignore those realities that do not fit preconceived conclusions. There is, obviously, a finite amount of liquid oil in the earth's crust, but our

FIGURE 4-3a

CUMULATIVE DISCOVERY CURVES AND EUR FOR THE SAN JOAQUIN BASIN

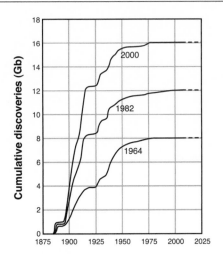

SOURCE: Based on graph in Nehring (2006a).

estimates of this grand total remain uncertain. Thus, Hubbert's value for the estimated ultimately recoverable (EUR) oil was too low because he had no knowledge of the Prudhoe Bay supergiant oilfield or of coming giant finds in the Gulf of Mexico. While our knowledge of the EUR is a constant work in progress, the maximum rates of oil extraction have been trending upward, from typical rates of less than 30 percent of all oil in place two generations ago to more than 40 percent in some reservoirs today.

Moreover, a much documented reality is that an oilfield's ultimate recovery tends to grow with time because of additional drilling and higher recovery rates; EUR oil for recently discovered fields thus definitely underestimates their eventual cumulative production. Nehring demonstrated how this reality invalidates predictions based on Hubbert's method for two of the leading oil-producing regions in California and in Texas and New Mexico.[14] This reserve growth tends to be substantial, not infrequently doubling the original estimate (see figures 4-3a and 4-3b). Thus, Nehring concludes, "The task facing us now is not to continue to use an obsolete and irrelevant method [that is, Hubbert's model] but to develop further our understanding of recovery growth."[15]

FIGURE 4-3b

CUMULATIVE DISCOVERY CURVES AND EUR FOR THE PERMIAN BASIN

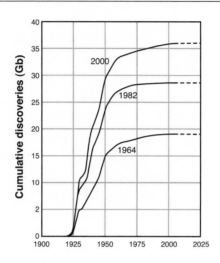

SOURCE: Based on graph in Nehring (2006b).

Advocates of an early oil peak make a very valid point about the absence of rigorous international standards in reporting oil reserves, and they also correctly point out that many official oil reserve estimates should not be trusted because they reflect political bias. And the considerable uncertainty surrounding global estimates of ultimate recovery means that no such figures can be taken as definite values to construct the symmetrical exhaustion curves. Or, as a life-long observer of oil affairs has put it, "To know ultimate reserves, we must first have ultimate knowledge." But nobody has this knowledge, and "nobody should pretend to."[16]

Untapped Resources

The latest assessment of global oil resources by the U.S. Geological Survey (USGS) concluded that nearly 690 billion barrels of oil could be added from the appreciation of currently known fields, and some 730 billion barrels are yet to be discovered.[17] This estimate puts the global ultimate recovery total at about 3.02 trillion barrels. And even the estimate whose probability the USGS puts at 95 percent (that is, a virtual certainty) adds up to about 400 billion barrels, or almost three times as much as a typical claim advanced by advocates of an imminent peak in oil production.

The largest new discoveries of conventional oil are expected in the Mesopotamian Foredeep Basin (extending from just north of Baghdad through Iraq and Kuwait to the Eastern Province of Saudi Arabia), in the West Siberian Basin, in the Zagros Fold Belt of southeastern Iran, in the Niger Delta, in the vast Rub al Khali ("Quarter of Emptiness") Basin of eastern Saudi Arabia, and in the as yet unexplored East Greenland Rift Basin. The best prospects in North America are in northern Alaska, in the Canadian Arctic, and in the Gulf of Mexico, while major reserve additions in Latin America should come in Venezuela and offshore Brazil. Africa's largest untapped oil resources are in waters off Congo and Niger, as well as in Algeria and Libya. Asia's brightest prospects are in Kazakhstan and the enormous Timan-Pechora Basin west of the Urals, and Europe's most promising area is the Atlantic margin west of Scotland.

The point is that until all of the world's major sedimentary basins have had the density of exploratory drilling comparable to that of Texas or Oklahoma, there is no compelling reason to favor the most conservative estimates of ultimate recovery preferred by the peak-oilers rather than the higher totals offered by other geologists. And I must stress that this comparable exploratory density should include all deeper offshore waters; after all, half a century ago there was no true offshore (out of the sight of land) oil extraction, and today's routine production in deep water was unthinkable twenty years ago.

In any event, for a truly realistic assessment we have to go beyond the uncertain EUR of conventional oil and adopt a broader resource perspective. Even Laherrère conceded that the addition of the median reserve estimates of natural gas liquids and nonconventional oil would double his EUR value. An assessment by Cambridge Energy Research Associates put the global oil resource base of conventional *and* nonconventional resources, including the historical cumulative production of 1.08 trillion barrels, at 4.82 trillion barrels and likely to grow.[18] This means that 3.74 trillion barrels remain to be extracted, and that the future of global oil production is best imagined as an undulating plateau rather than a steep decline mirroring the historical incline. Nonconventional oil resources will have a critical role in forming and extending this plateau—but their future contribution is itself the subject of myths and misconceptions, and we should not think that these (undoubtedly large) resources will easily make up for the declining output of many old giant oilfields.

Nonconventional Oil Reserves

It needs to be understood that no sharp line divides conventional and nonconventional oil resources. The continuum of hydrocarbons runs from medium-heavy oils that are mobile at reservoir conditions, to somewhat mobile extra-heavy oils, to tar sands and bitumen that are nonmobile within their reservoirs, to oil shales that have virtually no permeability. Heavy oils of medium and even very high densities have been extracted for decades in Saskatchewan and Venezuela as well as from Alaska's North Slope, where up to 40 billion barrels of this nonconventional fuel may be in Prudhoe and Kuparuk, the slope's two supergiant conventional fields. But as low temperatures and permafrost make the oil even more viscous, no more than 20 percent of this total is seen as potentially recoverable, and the current extraction accounts for just 5 percent of the North Slope output.

Similar nonconventional deposits are found in many oil basins around the world, but most of the 4–5 trillion barrels of heavy oils are in Venezuela—at least 1.2 trillion barrels, of which 270 billion barrels may eventually be recoverable—and in Alberta's oil sands in the province's Athabasca region, which contain some 2.5 trillion barrels of bitumen. Extraction of oil from Alberta's oil sands has been commercially viable for decades. Suncor began production near Fort McMurray in 1967, and the Syncrude consortium has been producing in the same area since 1978. Both companies operate large excavators in sprawling, opencast mines and use the world's largest off-road trucks to take the sand to extraction plants, after which the liquid goes to upgrading facilities that convert it to light, low-sulfur crude oil. Only about a fifth of recoverable oil in Alberta's oil sands can be reached by mining; the rest will have to be extracted *in situ*.

The two commercial techniques used in new operations are cyclic steam stimulation and steam-assisted gravity drainage. Cyclic steam stimulation, which was developed by Imperial Oil at its Cold Lake Project in

FIGURE 4-4

CYCLIC STEAM STIMULATION

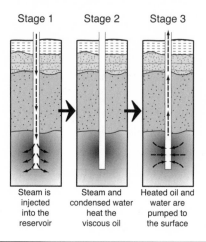

Stage 1	Stage 2	Stage 3
Steam is injected into the reservoir	Steam and condensed water heat the viscous oil	Heated oil and water are pumped to the surface

Source: Based on Liu (2006).

Alberta, injects hot pressurized steam into well bores and then leaves the steam to soak for a period ranging from a few months to three years to loosen the bitumen; the heated bitumen-water mixture is then pumped to the surface. As much as 35 percent of the bitumen originally in place can be extracted this way (see figure 4-4).

Steam-assisted gravity drainage uses two horizontal bores 500–800 meters long and separated by a vertical distance of 5 meters near the bottom of an oil sand formation. Steam injected into the top bore heats the surrounding bitumen and drains it into the bottom bore, with recovery rates of up to 60 percent. To produce the steam, both of these techniques require large volumes of water and energy (now mostly from natural gas), and their availability and price will be the key determinants of future extraction. A new extraction method under development combines vertical air injection with a horizontal production well to create a narrow combustion front that heats the bitumen and drains it into the well; its commercial success would largely preserve the surface vegetation, as well as greatly reduce the overall amounts of water and energy used to recover the oil from sands.[19] The latest assessment by the Canadian Association of Petroleum Producers put the total of Canada's conventional oil reserves at about 5.4 billion barrels as of the end of 2007, and it added to this 8.9 billion barrels of oil recoverable from the already developed commercial oil sands projects.[20] CAPP also put the total volume of oil that could be recovered by existing techniques at about 175 billion barrels, second only to Saudi Arabia's 264 billion barrels and well ahead of the reserves for Iran, Iraq, and Russia—respectively, 137.5 billion, 115 billion, and 79 billion barrels in 2008.[21]

But there is a huge gap between Alberta's enormous oil sand reserves and any realistic extraction rate in coming decades. Ignoring this reality has led to a myth in response to the peak oil myth that expanded oil sand extraction will be able to make up for most of the declining production now observed in many of the world's giant oilfields, but there is no doubt that this cannot happen. In 2005, Alberta's oil sands yielded 1 million barrels per day (Mbpd); by 2008 the total was up to 1.3 Mbpd; and the prospects are for 2 Mbpd by 2011. Plans to produce 3.5 Mbpd by the year 2015 have faltered as a result of the economic downturn in 2008 and 2009.

Annual extraction on the order of 3 Mbpd would not be insignificant— it would be about 20 percent more than the current output by Venezuela, and equal to the output of the United Arab Emirates—but it is clear that Alberta's oil sands, regardless of all the boasts about its resources being bigger than Saudi Arabia's, will not be a decisive factor in ensuring global oil supplies during the next ten to twenty years.

Future rates of nonconventional oil production will be determined by a complex interplay of oil prices, perceptions of supply security, and technical advances. Even if we had perfect knowledge of the world's ultimately recoverable conventional oil and of the realistically recoverable amounts of nonconventional oil resources, drawing the future production curve with a high degree of confidence would still be impossible, because we cannot know future oil demand; the economic downturn of 2008 and 2009, accompanied by a sharp drop in demand, has been just the latest proof of this inability. Future demand will be driven by predictable factors—such as growing populations and higher disposable incomes in Asia—as well as by unpredictable political and socioeconomic changes, and, most important, by new technical advances. Simply put, those who await an imminent peak in oil production ignore the fact that the shape of the global oil extraction curve is affected by price, a phenomenon clearly demonstrated by the decline and stagnation of global oil consumption brought about by high prices.

Production, Demand, and Prices

After OPEC nearly quintupled its oil price in 1973, the initial response was small. Global oil consumption declined by merely 1.5 percent in 1974, and by 1976 it was nearly 4 percent above the 1973 level. While oil use may have been inelastic to the initial quintupling of the world price, however, it subsequently responded to an additional near trebling that took place between 1978 and 1981. By 1983, the world's oil production had fallen by nearly 15 percent from the record level of 1979—an obvious sign of a vigorous market response, and not of a dwindling resource. The record 1979 extraction was not surpassed until fifteen years later, in 1994, although a 1973 forecast using the demand growth rate of the previous decade would have put the 1990 demand nearly three times higher.

The same market forces have been mobilized by the latest oil price rise that began in 2004–5. In 2006, oil demand was down in nearly all the leading affluent countries that imported oil. In 2007, U.S. consumption was down by 0.1 percent, while the declines were 3.5 percent in Japan and 5 percent in the United Kingdom. In 2008 the U.S. demand drop was more than 6 percent, U.K. demand remained flat, and Japanese demand was down by more than 3 percent.[22]

Lower demand, and not any imminent physical shortage of oil in the ground, was the main reason (besides OPEC's usual manipulative refusal to produce more) that global oil production was essentially flat in 2007 (down by 0.1 percent) and 2008 (up by 0.6 percent).

A growing share of the market claimed by alternative fuels would tend to postpone the arrival of global peak oil extraction, particularly as the acceptance of higher prices for alternative fuels may intensify the efforts to extract even more oil from known reservoirs; after all, even today's best recovery methods still leave behind 40–50 percent of oil originally in place. Future contributions of any conceivable alternatives will depend on many factors. Among the most notable are long-term changes in global demand and national and international commitments to technical innovation. Many forecasters now treat China's near-double-digit growth of gross domestic product as an inertial phenomenon that will continue for decades and draw in increasing amounts of oil. They might look closer at similar forecasts during the 1980s that assumed continuing ascent of the Japanese economy to the world's dominant place—and at the post-1990 Japanese realities. And a serious commitment to technical innovation may bring cheaper, non-oil-based fuels, superior car batteries, or less wasteful long-distance trans-mission of electricity, changes that may not be enough to displace most of the existing oil demand but that could significantly reduce oil intensities (amount of oil used per unit of GDP) of all major economies.

And even greater declines in oil demand—and hence a later peak oil date—could be brought about by aggressive resource management and preferential allocation to ensure the availability of refined products for such existential necessities as fueling agricultural machinery, maintaining basic airline connections, producing key feedstock for essential petrochemical syntheses, and transporting perishable goods. Technical advances unfold across decades, but their eventual impact on resource demand is profound. In 1930, before the invention of the jet engine, nobody predicted the emer-gence of large-scale commercial jet-based aviation, the industry that is now the leading consumer of kerosene. And in the early 1980s, as oil prices rose to record levels, nobody predicted that twenty-five years later half the vehi-cles in the United States would be gasoline-guzzling SUVs, pickup trucks, and vans. In contrast, a universal adoption of high-performance hybrid cars could halve the current demand for automotive fuel in two decades.

The combination of different estimates of ultimate oil recovery and dif-ferent rates of future oil demand results in many possible production curves, some of them extending far into the twenty-first century. For exam-ple, the combination of sustained technical advances, fuel substitutions, demand growth rates slower than in the recent past, and the global ultimate recovery of about 3 billion barrels would translate into the peak of conven-tional oil extraction coming sometime after 2020, with the global produc-tion during the 2040s possibly still as high as in the early 1980s. A vigorous expansion of nonconventional sources and alternative fuels could sustain a prosperous oil industry well after the year 2050 (see figure 4-5a). Peter Odell goes even further: In his acceptance speech for the Biennial OPEC Award for 2006, he concluded that today's peak-oilers, much like their

FIGURE 4-5a

POSSIBLE PEAKS OF OIL EXTRACTION

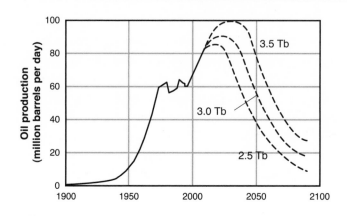

SOURCE: Smil 2003.

predecessors, will be soon proved wrong; that the present contribution of oil and natural gas to the global total primary energy supply will decline only modestly by 2050; and that while natural gas will become the leading fossil fuel, the oil industry at the end of the twenty-first century will still be larger than it was in 2000.[23]

A similar range of long-term outcomes was suggested at the 2006 Hedberg Research Conference organized by Richard Nehring. The best estimates offered by seventy-five experts from nineteen countries foresaw the earliest global production peak occurring at around 2020 with less than 95 Mbpd, a slightly higher extraction plateau starting at 2030 with a medium estimate of the ultimate recovery, and the maximum output of about 100 Mbpd, based on high ultimate recovery, reached by 2040; extraction in 2080 would still be above the 2005 level (see figure 4-5b).[24] As Abdalla Salem El-Badri, the OPEC secretary general, reiterated in a recent speech, "The issue is not whether the resources are there. We know they are. The world has enough oil resources to meet demand and satisfy consumers for decades to come."[25] His views are shared by Rex Tillerson, the CEO of ExxonMobil, the world's largest multinational oil company, as well as by Khalid Al-Falih, the CEO of the Saudi Aramco, the world's largest national oil corporation.[26]

FIGURE 4-5b

**POSSIBLE PEAKS OF OIL EXTRACTION CHARTED
BY HEDBERG RESEARCH CONFERENCE, 2006**

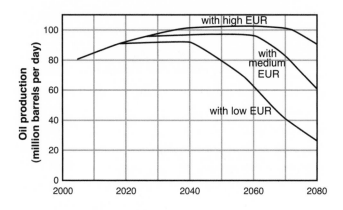

SOURCE: Based on Kerr (2007).

And yet what these men, who control most of the world's resources, claim is completely dismissed by many prominent peak-oilers as crass misinformation and biased propaganda; they believe that oil production in general, and the extraction from Saudi Arabia's giant oilfields in particular, is in a precipitous decline, and that no amount of technical innovation or number of future oil discoveries will make up for these extraction losses.[27] These peak-oilers are absolutely convinced that due to the fuel's physical shortages, peak global oil production has already been reached, or that it will arrive no later than by 2015 or, with luck, by 2020—and what awaits afterward is "life after the oil crash," as oil extraction goes into "terminal decline." These extreme positions regarding the future of the global oil supply (its imminent precipitous decline versus decades of continuing importance) cannot be reconciled.

But even if we were to assume an imminent peak of global oil extraction—or, as Simmons claims, a peak that has already taken place in 2005—there is no reason for espousing any end-of-civilization scenarios with declining oil extraction. A single revealing example illustrates why. Let us assume (rather dramatically) that in 2025 the total U.S. automotive gasoline supply will be 20 percent lower than it was in 2008. The Census Bureau forecasts the 2025 population to be 18 percent larger than in 2008,[28] which means that the per-capita supply of gasoline (and hence the average personal mobility) will be only 68 percent that of 2008—a dramatic drop of nearly one-third in just seventeen years. But after the intervening efficiency gains—with the CAFE average now mandated to rise by nearly 30 percent already by 2016,[29] with its further improvements after 2016, and with increasing numbers of hybrid vehicles on the road—it is very realistic to expect at least a 40 percent efficiency improvement in the use of America's automotive gasoline by 2025. As a result, the per-capita level of useful energy services provided by gasoline would be only 5 percent lower in 2025 than it was in 2008—a marginal decrease that could be easily accommodated by forgoing one trip out of every twenty—and most certainly not any harbinger of returning to an American version of the Olduvai Gorge! And, obviously, a slightly higher, but still very realistic, 45 percent efficiency gain would result in no change to useful services derived from a much lower gasoline consumption.

Countering the Claims of Peak-Oilers

As I have argued, however, conventional oil resources may, in fact, be substantially larger than the lowest estimates favored by peak-oilers; nonconventional ones are definitely abundant, although the rate of their future extraction may be relatively modest; Hubbert's symmetric production template does not fit most of the real-world cases, and the peak, however defined, is most likely to be followed by an extended plateau rather than by a precipitous fall; supply alternatives, above all natural gas and nonconventional oil, are available or can be developed over time; and demand can be lowered drastically, whether by concerted action or by a prolonged global economic slowdown.

Let us counter the claims of radical peak-oilers (those who see the peaking oil output as the beginning of civilization's dramatic collapse) calmly. Extraction of any mineral resource must eventually decline and cease, whether due to actual physical exhaustion of a particular deposit or, much more commonly, for economic reasons, as the rising financial cost and falling net energy return force the use of alternatives. Obviously, conventional crude oil will not be an exception. It is fairly probable that its annual global extraction will peak within the next two decades, and it is inevitable that its share of the world's primary energy supply will continue to decline. In 1980 oil provided 44 percent of the global primary energy supply, by 2000 it was down to 41 percent, and in 2009 it stood at less than 35 percent (though its absolute extraction in 2008 was nearly 32 percent above the 1980 level).[30]

In any case, the fuel's declining share of the global commercial primary energy supply spells no imminent end of the oil era; given the very large remaining conventional and nonconventional resources, oil will continue as a major contributor to the world market during the first half of the twenty-first century. As it becomes dearer, we will use it more selectively and more efficiently, and we will intensify a shift that has been underway during the past generation: a new global energy transition, from oil to natural gas and to both renewable and nuclear alternatives, with the latter two options having a potential to capture significant (but not dominant) portions of the global energy supply by 2050 and displace notable shares of oil-derived fluids in some nations.

As a result, there is nothing inevitable about any particular date of peak global oil extraction; more fundamentally, there is no reason to see an eventual decline in oil's share in the global energy supply as a marker of modern civilization's demise and, even more dramatically, as the beginning of humanity's return to the Olduvai Gorge. Energy transitions—from biomass to coal, from coal to oil and gas, from direct use of fuels to electricity—have always stimulated human inventiveness. They challenge both producers and consumers, necessitate the scrapping or reorganization of extensive infrastructures, are costly and protracted, and cause major socioeconomic dislocations. But energy transitions have also created more productive and richer economies, and modern societies will not collapse just because we face yet another of these grand transformations.

Paradoxically, our very wastefulness (itself a function of decades of inexpensive supply) is a major factor working in our favor. As I have argued,[31] and as the 2000-Watt Society project of the Swiss Federal Institute of Technology (ETH) tries to demonstrate,[32] a decent quality of life in the world's affluent countries could be secured even if they were eventually to halve today's energy demand. Unless we believe, preposterously, that human inventiveness and adaptability will cease the year the world reaches the peak annual output of conventional crude oil, we should see that milestone—whenever it comes—as a challenging opportunity rather than as a reason for cult-like thinking and paralyzing anxiety.

5

Sequestration of Carbon Dioxide

During the first years of the new millennium, it became clear that, among affluent countries emitting large amounts of carbon dioxide, some would not only fail to meet the relatively undemanding targets for reducing CO_2 emissions specified by the Kyoto Protocol, but would even fail to stabilize CO_2 emissions at their current, very high, per-capita levels. Most notably, by 2005 the United States and Canada were, respectively, more than 20 percent and about 55 percent above their 1990 emissions.[1] And the emissions growth was much faster in the two leading economies that were not included in the reduction targets of the Kyoto Protocol, those of China and India. China's rising coal combustion and expanding oil imports were the most important drivers of this process. In 2005, China's overall CO_2 emissions were roughly 2.15 times higher than in 1990, and, in 2006, it surpassed the United States to become the world's largest emitter of CO_2. In India, emissions doubled between 1990 and 2005. Underlying trends in the economies of these countries could sustain similarly high growth rates for at least a generation. Consequently, all the politically correct talk about sustainability, energy conservation, and the "greening" of economies has not acknowledged what is really taking place.

Barring an encounter with an extraterrestrial object or an unprecedented viral pandemic, global emissions of CO_2 are set to rise substantially during the coming decades. Even an unusually deep and prolonged worldwide economic downturn would merely interrupt or moderate, but not reverse, this trend. Currently, there is no single approach—that is, a technical process or a managerial or economic tool—that can prevent the doubling of the preindustrial CO_2 concentrations (about 270 ppm in 1850) and thus keep them below 450 ppm, a level that would (as best as we can tell) result in only a tolerable increase of average tropospheric temperature, perhaps no more than about 2°C. Because the growth of emissions appears inevitable, the focus of stabilization efforts has shifted to the sequestration—that is, the capture and removal—of the emitted CO_2, and the paths to sequestration include both enhanced natural processes and new engineered methods. Hopes may be high, but it is highly unlikely that during the next few decades, carbon sequestration will remove enough CO_2 to arrest, or even to slow down substantially, the atmospheric accumulation of the gas.

Organic Approaches

Earth's biosphere is, of course, nothing but an enormous carbon sequestration/regeneration system. Photosynthesis uses water and CO_2 to produce new plant matter (about 45 percent of which is carbon), while decomposition of organic matter returns the gas to the atmosphere (see figure 5-1). Annually this process is responsible for moving more than 120 billion tons of carbon from the atmosphere to plants, but plant (autotrophic) respiration promptly returns about half this amount to the atmosphere, leaving about 65 billion tons of carbon in a new plant mass that is then consumed by organisms ranging from bacteria and fungi (the leading agents of decomposition) to humans, and eventually returned to the atmosphere through heterotrophic respiration. There has been no scientific consensus about the recent global level of net carbon sequestration in plant mass.

Higher atmospheric CO_2 levels mean that plants have been storing annually 1.2 billion to 2.6 billion tons more carbon than during the preindustrial era[2]—but Potter and others found that, as a whole, the terrestrial biosphere can fluctuate widely, being a source of carbon in some years and its sink in others.[3] Continental and regional balances are similarly uncertain. North American vegetation appears to be a consistent, and a fairly large, carbon sink.[4] Eurasian forests also store carbon,[5] and this storage is increasing because of the continent's forest expansion and good management;[6] the net uptake of the Russian boreal forests is perhaps increasing as well.[7] Overall, the northern lands are undoubtedly a carbon sink, and even the tropical forests are absorbing more carbon than was previously thought.[8] But it would hardly be a surprise if new studies were to shift some of those rates.

FIGURE 5-1

EARTH'S BIOSPHERE: A CARBON SEQUESTRATION/REGENERATION SYSTEM

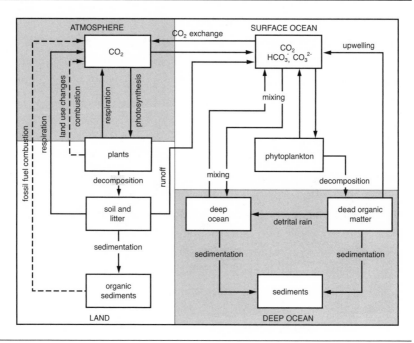

SOURCE: Smil 2000.

Global warming will make all these uncertainties even greater because it will lengthen the growing seasons and intensify water cycling—that is, the overall amount of precipitation will increase—in many regions. This combination will result in higher plant productivity, a trend that was already evident throughout most of the United States during the latter half of the twentieth century.[9] But what the long-term effect of such changes will be is not clear. Will the additional productivity be promptly negated by higher rates of respiration in a warmer world? Will most of its increment be stored in long-lived tissues, such as trunks and major roots, or tissues with rapid turnover, such as foliage and fine roots? And, most fundamentally, will global warming eventually convert forests from carbon sinks to carbon sources?[10]

Moreover, in the biosphere's largest remaining tropical rain forests, in the Amazon and Congo basins and in Borneo, future net carbon fluxes will be determined primarily by the rate of deforestation, while the capacity of many forests to sequester carbon will be limited by water and nutrient constraints, particularly by the availability of nitrogen.[11] This is why, in the absence of further specific land management, the current European net carbon uptake is expected to decline soon.[12] And in some ecosystems, enhanced carbon storage could be outweighed by carbon losses through a higher frequency of regional wildfires and longer duration of droughts in a warmer world.

New Plantings. Most of these uncertainties would not come into play with new plantings; extensive plantations of fast-growing trees, afforestation of areas that are currently barren but receive enough precipitation and have good enough soils to support new mixed-tree plantings, and the restoration of destroyed climax vegetation would indisputably result in a considerable storage of carbon. The two fundamental problems with this approach are the durability of such carbon sequestrations and the scale of intervention needed to make a real difference.

Fast-growing species may reach maturity in just ten to fifteen years, after which they will have no (or just negligible) net carbon storage. Longer-lived plantings (including temperate and boreal communities of deciduous or coniferous species or their mixtures) will mature only in forty to eighty years, but the rate of their carbon storage slows down as they approach maturity. Many of them, moreover, may succumb prematurely to pests or fires, or their growth can be retarded by droughts or cyclone damage. Consequently, the additional carbon sequestration achieved by planting more trees is either rather short-lived or its long-term contribution is difficult to quantify with much certainty. In any case, very large areas of new plantings would be needed to offset a significant share of CO_2 emissions. Sequestering just 10 percent of the 2005 total (about 800 million tons of carbon a year) would require an area of boreal plantings nearly as large as the combined area of forests in North America and Russia, or the equivalent of nearly 15 percent of the total area of today's tropical rain forests.[13]

Soil Organic Carbon. The long-term fate of soil carbon is even more uncertain than the storage of tree carbon and the rates of storage in new plantings. Soils already store more than twice as much carbon as is in the atmosphere and nearly four times as much as is in the terrestrial plants. Appropriate agronomic management—conservation tillage, the planting of cover crops, and crop rotation—could significantly enhance the rate of additional storage in the form of soil organic carbon and thus not only help lower atmospheric CO_2 levels but also improve soil productivity.[14] Potential storage capacity is large because in many intensively farmed regions, the levels of soil organic carbon are half of preagricultural levels. At the same time, increasing tropospheric ozone levels can reduce plant productivity and hence also significantly lower soil carbon formation.[15]

Global warming will complicate attempts to generalize and make predictions about soil carbon because it will accelerate decomposition and cause additional releases of CO_2 from soils in a process that could reinforce the warming. This response has been confirmed not only by models and small-scale experiments but also by some large-scale analyses.[16] In addition, rising CO_2 levels have been shown to reduce the sequestration of root-derived carbon in the soil for many European tree species.[17] But it would be premature to extrapolate these results worldwide, because few natural processes are as complex and as intricately interactive as those affecting the fate of organic carbon in soils.[18]

As a result, we cannot predict even the direction of soil carbon change. Will the carbon released by warming-driven acceleration of soil respiration create a significant positive feedback, or will a significant share of additional photosynthesis be stored in long-lived soil organic matter and act as a notable carbon sink? We need to bear in mind, as well, that additional carbon storage in temperate ecosystems may be negated by rising emissions of methane, a much more powerful greenhouse gas than CO_2, from warmer sub-Arctic and Arctic wetlands and lakes; in one Siberian region, its emissions may be already five times higher than previously estimated.[19]

Biochar. The latest effort to increase carbon sequestration involving terrestrial plants is the incorporation of charcoal into soils. This method was inspired by Amazonian *terra preta* soils, whose high productivity is largely explained by high concentrations of charred biomass. Biochar, as it is now called in soil science literature, can hold two to two and a half times as much carbon as soils from the same parent rock that lack it. It also provides exceptionally long-lived storage of the element—at least several hundred and possibly even a thousand years—and its unusually high adsorption (activated charcoal in filters is just a cleaner version of this material) also binds a great deal of soil organic carbon.

These characteristics have led to proposals for an unorthodox use of biomass as a means of carbon sequestration.[20] Pyrolysis of plant tissues (low-temperature heating without oxygen) would produce charcoal that would retain about half the carbon that was initially present in the biomass, compared to just 10–20 percent retained by a decomposed biomass, as well as a gas. The gas—mainly CO with a bit of H_2 and CH_4—would be used as a source of energy, and the charcoal would be incorporated into soils. Crop residues and forestry waste would be the obvious candidates for pyrolysis, and fast-growing species could be planted on suitable land to be charred. Lehmann, Gaunt, and Rondon have gone so far as to claim that by the year 2100, biochar sequestration could store more carbon than is emitted from the combustion of fossil fuels.[21]

There are many reasons why this will not happen. To avoid crippling rates of wind and water soil erosion, a large part of the residual crop phytomass must be recycled, and whatever can be safely removed is now touted as a feedstock for the future cellulosic ethanol industry, raising the potential competition for straw and other residual phytomass. We have no idea what might be the eventual breakdown of these three uses (protecting against erosion, pyrolysis, conversion to ethanol). Moreover, because biochar requires arable agriculture to be incorporated into soils, it would be at odds with efforts to reduce soil erosion through no-till farming. Any large-scale pyrolysis of crop residues or forestry wastes would also involve a logistical challenge, particularly in mountainous terrain, since large fleets of mobile pyrolyzers would have to follow harvest or logging crews and convert the available residues. And even if logistical problems were solved, the amount of carbon trapped by pyrolysis would be very small. A simple calculation shows this: Affluent countries now produce annually about 900 million tons of dry straw. If all of it were pyrolyzed, the carbon sequestered would amount to only about 2.5 percent of 2005 global CO_2 emissions from fossil fuel combustion. In any case, the basic assumption is unrealistic, as a large part of that straw should always be used first to reduce erosion and to provide valuable animal roughage feed and bedding.

Enhanced Phytoplankton Production. Carbon sequestration through enhanced phytoplankton production is an even more uncertain, and a logistically forbidding, proposal. Its efficacy rests on the fact that iron controls the productivity of nutrient-poor waters of the open ocean.[22] Adding iron to the water can double the specific growth rate of phytoplankton and increase its abundance by an order of magnitude in a matter of days or weeks. This drawdown of CO_2 in the surface water layer is usually followed by zooplankton blooms, whose fecal pellets and dead bodies produce enhanced transfer of organic carbon to the deep ocean. The efficacy of this carbon pump has been confirmed by experiments in both equatorial and southern ocean waters.[23] Because roughly 20 percent of the ocean's surface has relatively high nitrate levels but very low iron concentration, fertilization with iron has been proposed as an effective means of sequestering carbon from the surface ocean. After the planktonic algae whose growth was stimulated by iron died, their dead cells would sink to the ocean bottom and become a part of sediments that can store carbon for millions of years.

The problem with this approach is that higher surface productivity may not translate into a commensurately higher removal of carbon into the abyss. While equatorial experiments showed that the iron stimulation works, the carbon pump was not activated by iron additions in colder waters south of Australia. In addition, the effects of continuous massive additions of iron to surface seas are not known; they might stimulate toxic algal blooms rather than the desired detrital rain to the abyss, and much of the new algal biomass might be promptly respired. This uncertainty would dictate a very cautious approach. According to Tréguer and Pondaven, moreover, it might be silica rather than iron that is the ultimate controller of CO_2 sequestered by ocean phytoplankton.[24] And the largest experiment with iron enrichment, the Indo-German fertilization of 300 km^2 in the southwestern Atlantic in March and April 2009, resulted in enhanced production of phytoplankton species that, unlike silica-coated diatoms,were not protected by hard shells and were hence avidly eaten by amphipods—small crustacean shrimp-like zooplankton—which prevented further growth of iron-stimulated bloom.[25]

In any case, the logistics of continuous ocean fertilization would be demanding, as they would require a large fleet of vessels incessantly applying fine particulate iron over large areas of the open ocean, often in extremely inclement weather. Legal aspects of such large-scale and continuous inter-ventions would also require new international agreements, and hence their early massive deployment is unlikely. I agree with a recent survey of ocean fertilization experiments: "Adding iron to the ocean is not an effective way to fight climate change, and we don't need further research to establish that."[26]

It is impossible to make any definite judgment about the rates and mag-nitudes of future carbon sequestration in natural ecosystems or in newly afforested (or reforested) areas, or about carbon storage gains resulting from better management of forests, grasslands, and soils. The best evidence indi-cates that the carbon sequestration potential of croplands has most likely been overestimated,[27] while the storage capacity of old forests has almost certainly been significantly underestimated.[28] Neither of these conclusions makes it any easier to design and implement any long-range sequestration programs with clearly achievable goals. And the constancy and scale of the requisite activity make iron-driven carbon sequestration into the abyss an even more dubious enterprise.

Technical Fixes

This leaves us with technical fixes that do not involve living organisms and whose outcomes can be much more closely determined and controlled. They have included proposals ranging from theoretically intriguing but practically doomed ideas to methods that could be made to work but whose efficacy is limited and whose cost is high.

The Nuclear Solution. Two theoretically intriguing examples include Mar-chetti's nuclear solution and mass storage in Indian basalts. Marchetti pro-posed to solve the CO_2 problem "without tears" by using natural gas as the principal fossil fuel, subjecting it to steam reforming using high-temperature nuclear reactors, and then reinjecting CO_2 into the original gas fields.[29] Even if the entire exercise were to be profitable, the enormous infrastructure needed for it to work—hundreds of new nuclear reactors of a type that has yet to be widely commercialized (including in the countries that oppose new construction of any nuclear plants) and an extensive network of new CO_2-transporting pipelines—would take many decades to build.

Storage in Basalts. Proposals to capture CO_2 within and below the basalt layers of India's extensive Deccan Traps ignore the facts that the basalts are not very porous and that they are already highly weathered as well as hot and highly fractured, making it unlikely that the gas would stay confined.[30] A similar scheme involves the storage of America's CO_2 emissions in the permeable undersea basalts (more than 1.6 miles below the surface) of the Juan de Fuca tectonic plate just off Seattle and Vancouver.[31] I assume that few investors would be eager to build a 3,000-mile pipeline from the East Coast to carry those emissions to the Pacific seafloor. But even if those basalts were to sequester all CO_2 from large stationary sources in the three Pacific states, such a scheme would store away only about 4 percent of all U.S. CO_2 emissions. Obviously, we could prevent much larger emissions merely by mandating higher CAFE standards for SUVs.

Mineral Carbonation. Yet another unorthodox proposal is for the absorption of CO_2 by the weathering (carbonation) of exposed peridotite, a mineral found in the Omani desert with high affinity for the gas;[32] this process could be accelerated by drilling or hydraulic fracture. But even leaving aside the practicalities of the actual storage procedures, the passage of CO_2-carrying tankers from the United States and China to Oman is hardly likely. Furthermore, a preliminary assessment suggests that roughly 1 billion tons of the gas (most logically piped in from the nearby Persian Gulf region) could eventually be sequestered annually. While that is a substantial amount, it is still equivalent to storing just two months' worth of China's current CO_2 generation.

Carbon sequestration using mineral carbonation and CO_2 removal directly from the air are two other ideas that are unlikely to be commercialized in the years ahead. Carbonation has at least three major advantages: The reaction generates a great deal of heat; it can proceed at low temperatures; and the mineral feedstocks required for it (silicate rocks) are quite abundant. Moreover, the resulting products (calcium and magnesium carbonates) are nontoxic solids suitable for simple aboveground disposal by landfilling or storage at surface mine sites. But the mass of required reactants is very large, and it is very unlikely that this process will make a meaningful difference. I will illustrate this using magnesium rocks.

The minimum requirement for carbonation with magnesium oxide ($MgO + CO_2 = MgCO_3$) is 0.9 tons of the oxide to remove one ton of CO_2, but in practice the process would use one of the abundant magnesium silicate minerals, such as forsterite or serpentinite. The minimum requirement for CO_2 reaction with serpentinite—$Mg_3Si_2O_5 (OH)_4$—is 2.1 tons of the rock to remove one ton of CO_2, but partial ore recovery and incomplete conversion during the carbonate reaction raise the total to at least three tons of ore for every ton of sequestered CO_2. Even assuming that this kind of sequestration would be done just for CO_2 produced from coal combustion—thus controlling only about a third of all anthropogenic emissions, which would not be enough to prevent a worrisome increase of atmospheric CO_2—the mining effort would be staggering.

In 2005, about 12 billion tons of CO_2 were released from coal combustion, and hence the annual extraction of serpentine ore needed for sequestration would have to surpass 33 billion tons, an amount nearly three times as large as the combined mass of all fossil fuels (less than 12 billion tons) extracted in the same year. For this reason alone, and leaving entirely aside the costs of this extraction and the energy required to mine and to move the ore to sequestration sites or to transport the gas to mine sites, this option will remain on paper. Even the greatest determination and unprecedented financial sacrifice could not create an industry that would handle several times the mass of global fossil fuel extraction, and do so before 2025, to keep the atmospheric CO_2 within an acceptable range.

Extraction from the Air. Similar considerations apply to the extraction of CO_2 from the air.[33] This process could take place in tall metal towers; as the air flows through the structure, CO_2 would be absorbed either by a liquid sorbent sprayed in a fine mist or by thin sheets of an alkaline compound emplaced in the structure. Lackner estimates that a single "synthetic tree" could remove 90,000 tons of CO_2 per year, and hence about 160,000 such structures would be needed to capture half of the CO_2 emissions from the 2005 combustion of all fossil fuels. To build that number of capture towers would clearly be a manageable structural challenge. But the process itself, as appealing as it sounds—Lackner even touts backyard TV-size units that could remove 25 tons of CO_2 per year, or roughly the average U.S. per-capita production—requires solving several engineering challenges.

Since low wind speeds near the ground (and frequent calms in many regions) would limit the air throughput, faster flow rates would have to be created artificially or by siting the structures at windy locations—though such places may not be the best sites for storing the gas. Constant transportation of the sorbent to the contact surfaces would not only be a design challenge, especially in high winds; it would also be highly energy intensive. An aqueous solution of calcium hydroxide—$Ca(OH)_2$, or slaked lime—readily absorbs CO_2 by forming $CaCO_3$, but high temperatures (and, obviously, a constant source of energy) would be needed to recover the sorbent from this tightly bound carbonate in a slurry. And afterward there still would be billions of tons of gas to compress into liquid form and dispose of underground. Costs of this transportation and storage cannot be reliably estimated until this kind of capture graduates from theoretical musings to large-scale, continuously running operations—and none of the latter is about to become a commercial reality. Air capture of CO_2, then, is obviously yet another intriguing design that has a very low chance to be a part of any meaningful effort to limit the levels of atmospheric CO_2 during the coming generation.

Large-Scale Industrial Carbon Capture and Sequestration. The solution that has received most of the recent attention—capturing CO_2 from its concentrated combustion sources and storing it for the long term where it cannot easily reenter the atmosphere—is certainly much more practical. Proponents of carbon capture and sequestration (CCS) rightly claim that every one of its key components is a well-established engineering practice. Scrubbing of CO_2 from natural gas and hydrogen with aqueous amine has been done commercially since the 1930s.[34] Separation of CO_2 from flue gas emitted by coal-fired power plants would thus be an extension of a well-proven chemical process. Transporting the captured gas by pipelines would be another extension of an everyday practice: In the United States, nearly 4,000 miles of pipelines deliver CO_2 used in enhanced oil recovery (above all in Texas). Moreover, there would be no insurmountable technical problems to building five, or ten, times as many lines in a decade; after all, more than 70,000 miles of natural gas pipelines were built in the United States during the 1960s and again during the 1970s.[35]

And while the CO_2 that is currently used for enhanced oil recovery is not pumped underground with the objective of keeping it there permanently, the process of mass underground storage would differ from this widely used practice only in its choice of reservoirs. CO_2 would be stored in deep saline formations, in layers of rock permeated with brine, exhausted hydrocarbon reservoirs, or coal seams unsuitable for mining. There is no shortage of such structures, they are fairly widely distributed, some of them have enormous storage capacities, and it would not be prohibitively expensive to drill the wells that would be used to fill them. North America's deep saline formations alone could accommodate more than 1 trillion tons of CO_2 and could thus store more than a century's worth of the current U.S. emissions.[36]

Given that individual components of industrial carbon capture and sequestration are readily available, and that the underground storages have capacities sufficient for any volumes of CO_2 that might conceivably be captured during the twenty-first century, it is not surprising that this technical solution has been enthusiastically endorsed by large oil and gas companies— enterprises with the requisite expertise that would benefit tremendously by creating and operating a new industry. It has also found favor with governments and, as an explosion of publications indicates (Web of Science listed more than 5,000 papers on carbon sequestration by the end of 2009), it has become a very attractive topic for academics.

Let us consider in some basic detail what would be involved in actually implementing audacious plans for large-scale CCS. By far the most important consideration before launching any globally significant sequestration program is the immense challenge of scale—that is, the masses and volumes of the handled gas and the material and energy requirements for its gathering, compression, transportation, and underground storage. These requirements are best illustrated by contrasting them with those of existing sequestration activities.

In 2009, there were only three experimental CCS projects that had lasted at least five years and operated at an annual rate of 1 million tons or more.[37] The oldest one, in the Sleipner West field in the Norwegian North Sea, has been injecting 1 million tons a year into a saline formation since 1996. Since 2000, CO_2 for enhanced oil recovery and storage in the Weyburn oilfield in Saskatchewan has come via a 320 km pipeline from a coal gasification plant in Beulah, North Dakota; alternating volumes of the gas and water are injected at 1.5 km below the ground, and the gas pumped to the surface with oil is separated and reinjected.[38] This project also has a capacity of about 1 million tons of CO_2 per year. At In Salah gas field in Algeria, the excess concentration of CO_2 has been removed from natural gas and injected (at an annual rate of 1.2 million tons) into a brine formation 2 km below the surface since the year 2004. The total planned storage for each of these projects is 17–20 million tons.

In contrast, worldwide fossil fuel combustion generated about 32 billion tons of CO_2 in 2008, and the aggregate emissions will amount to more than 500 billion tons between 2010 and 2025. That mass is four orders of magnitude greater than the eventual overall storage capacity of the three experimental projects. About 60 percent of this volume originates in large stationary sources (electricity generating plants, cement production, refineries, and various industrial enterprises, above all iron and steel mills and petrochemical syntheses), and hence the volume available for centralized sequestration will amount to at least 300 billion tons between 2010 and 2025.

If a serious commitment to large-scale sequestration were to begin with just 15 percent of CO_2 emitted in 2008 (or about a quarter of all emissions from large stationary sources), a new industry would have to capture, transfer, and store about 4.8 billion tons of CO_2 a year. Handling it under atmospheric pressure would entail impractically large volumes, which is why the gas, with a specific density of 1.967 kg/m³, is compressed for more economic transport and storage. Compressed supercritical gas (behaving as a liquid) has a minimum density of 468 kg/m³ (less than half that of water), so 4.8 billion tons of CO_2 would then occupy the volume of roughly 10.2 billion m³ (1 / 0.468 = 2.136; 2.136 x 4.8 = 10.25). Further compression reduces the gas volume and requires relatively little additional energy—but because increasing temperatures (with increasing storage depth) would reduce the gas density faster than the increasing compression would elevate it, the gas is not injected at densities higher than about 800 kg/m³. Even if all gas were handled at that density, it would occupy some 6 billion m³. For comparison, global crude oil extraction in the year 2008 amounted to 3.93 billion tons, and, with an average oil density of 0.85 g/cm³, its volume was roughly 4.6 billion m³.

Consequently, even if we were to start with a modest goal of sequestering just 15 percent of all 2008 CO_2 emissions, we would have to put in place a gathering, compression, transportation, and storage industry whose annual volume throughput would be (depending on the stored gas density) 1.3–2.2 times that of the annual volume throughput of the world's crude oil industry, with its immense networks of wells, pipelines, compressor stations, tankers, and above- and underground storages. Like the infrastructure for the global oil industry, worldwide infrastructure able to handle 6 billion–10 billion m^3 of CO_2 every year could be put in place only over a period of several decades and at a cost that cannot be as yet satisfactorily estimated—and more than two-thirds of all CO_2 emissions would remain uncontrolled.

The stored gas could occupy considerably smaller volumes only if it were deposited under high pressure and low temperature in engineered structures or within shallow sediments at the bottom of the ocean, but such storage would entail many other uncertainties (not the least the numerous implications it would have with respect to the international law of the sea) and technical challenges yet to be solved. And the facts that plenty of underground storage capacity appears to be available on land and that we have built thousands of miles of CO_2 pipelines do not mean that all large sources of the gas are near a suitable storage. Even if they are, the choice of many possible storages may run into vigorous NIMBY objections, as would the routing of many CO_2 pipelines in densely populated areas (nearly all of the existing U.S. CO_2 lines are in rural Texas, New Mexico, Colorado, and Wyoming). Case-by-case selection, impact statements, and right-of-way negotiations, not theoretical gross capacity or average length calculations, would determine the eventual progress.

And CCS would not be cheap. Today's best estimates put the overall cost of capture and compression at $30–$75/ton of CO_2.[39] The expected cost is highest for plants burning pulverized coal and lowest for plants using integrated gasification combined cycle (IGCC) technology.[40] Transport, whose costs are highly dependent on the mass flow rate—with unit costs per fixed distance dropping exponentially with the rising annual throughput—could add as little as $1/ton CO_2 or as much as $10/ton CO_2 for a 100 km line.[41] The cost of storage, including the obligatory subsequent monitoring, would be highly site specific, depending as it would on the accessibility, depth, and porosity of target underground strata; today's best estimates have much the same range as the cost of pipeline transport—that is, roughly $1–$10/ton CO_2 injected.

Taking approximately $60/t as an overall conservative average would mean that global carbon capture and storage with an annual capacity of 4.8 billion tons—that is, sequestering just 15 percent of today's emissions, or about 25 percent of the flux from large stationary sources—would cost close to $300 billion a year. But getting to the point where capture and storage were possible would first require a large capital investment. Recent estimates indicate that for pulverized coal plants with carbon capture and storage, the cost per installed kW would be about 60 percent higher than for plants without capture, and for IGCC, the difference would be at least 30 percent. But all of these values are highly uncertain.

While we can specify the technical requirements of today's large-scale sequestration projects, we can have only moderate confidence in estimates of an endeavor that would eventually have to be scaled by three to four orders of magnitude compared to today's experimental processes. Moreover, we must avoid a now common mistake that has been induced by rapid declines in unit prices for electronic goods. The production, and in part the design, of advanced microprocessors is an entirely automated process that needs low labor and material inputs and is highly conducive to falling unit costs with mass output. In contrast, large assemblies of the requisite infrastructure for carbon capture and storage (capture plants, pipelines, compressors, injection sites) will have to be tailored to specific conditions, their construction and maintenance will be highly labor intensive, and the materials needed by a massive new sequestration industry would put further pressure on the steadily rising costs of steel, aluminum, plastics, and concrete. Consequently, we cannot exclude increased, rather than decreased, unit costs with a future mass adoption of carbon capture and storage. Whatever the eventual case, the costs will not be trivial, and, except for the limited volumes of CO_2 used in the enhanced oil recovery, they will represent a net expense to all participating industries.

The Energy Penalty on Sequestration

The energy penalty exacted by capturing and storing fossil carbon is also uncertain. Large, coal-fired electricity generating plants are the best targets for the capture and compression of CO_2. But they would increase their internal electricity consumption—usually amounting to less than 10 percent, mainly for particulate removal by electrostatic precipitators and for flue gas desulfurization—by at least 30–40 percent. The carbon emissions per unit of generated electricity would thereby increase and reduce the net amount of the captured CO_2. The capital costs of an entire full-scale system serving a major electricity generating plant cannot even be estimated with much certainty.

A special report by the Intergovernmental Panel on Climate Change (IPCC, the organization that issues periodical consensus review about the progress of global warming) estimated that the cost of capture alone would raise the capital expenditure of a new pulverized coal–fired power plant by anywhere between 44 and 74 percent, and that the upper bound for the cost of capture and underground storage might nearly double the overall investment.[42] Even for a more efficient natural gas–fired station with a combined cycle (that uses the exhaust heat to generate additional electricity), the overall cost might be up to about 80 percent higher. These very substantial penalties would significantly reduce the net energy return on electricity generation equipped with carbon capture and storage.

And concerns and uncertainties would not end after the gas was sequestered. Highly site-specific conditions preclude any generalizations, either about the rate of future slow leakage or the probabilities of sudden discharge, which could be caused by the migration of gas to a preexisting but unknown fault or fracture or by an earthquake. The storage itself may be a source of future problems. An experimental injection of CO_2 into a brine reservoir in Texas at a depth of 1,500 meters was followed by a sharp drop in pH and dissolution of carbonates, a process that could ultimately open pathways in the rock seals or in well cements for CO_2 leakage.[43] Moreover, dissolution of iron oxyhydroxides due to the increased acidity could also mobilize toxic trace metals, as well as toxic organic compounds that could migrate into aquifers tapped for drinking water.

The best outcome of long-term CO_2 storage would be the precipitation as carbonate minerals. But a study of nine natural gas fields (the best natural analogs for appraising millennia-long storage of CO_2) found that only a small fraction of the emplaced CO_2 would be bound in that fashion, and that most of it would actually dissolve in the surrounding water.[44] This would obviously increase the possibilities of future leakage of CO_2-laden waters. For these reasons, it would be best to store CO_2 in deep aquifers whose impermeable shale caps would prevent any major leaks. But with a truly massive global storage, even tiny leaks could add up to significant overall rates after fifty to one hundred years, when the aggregates of tens (or even hundreds) of billions of tons would have been sequestered; an annual leakage rate of a mere 0.1 percent could amount to 0.5 billion–1 billion tons of carbon, a significant share of the limited mass of emissions that would effect a further rise in atmospheric CO_2. The health impacts of leaks should not be a major worry, but they cannot be ignored, either. CO_2 is innocuous in normal atmospheric levels (now 0.038 percent) and has no adverse health effects in levels up to 0.5 percent, which is the U.S. occupational exposure limit for eight hours, but it is lethal (asphyxiating) in high concentrations.

Consequently, it must be expected that the location and certification of CO_2 storage sites would encounter arguments, controversies, and resistance not unlike those encountered by the siting and operation of other hazardous facilities. Citizens Against CO_2 Sequestration is already organizing on the Internet,[45] and local opposition groups have been emerging in the United States, Sweden, and Germany; and as for the problem with the siting of major storages, it might be salutary to recall the decades-long opposition that has helped to delay the commissioning of Yucca Mountain, America's first permanent depository of highly radioactive wastes. Low-probability but potentially deadly risks would have to be taken into account,[46] and new national and international regulations would have to be put in place to deal with unprecedented contingencies.[47] Clearly, even if major steps were taken soon to move carbon capture and storage beyond small-scale trials to routine large-scale applications, several decades would have to elapse before the process made a significant dent in still-rising CO_2 emissions.

But there are no signs of a commitment to taking such steps. The principal promoters of large-scale sequestration—besides the academics eager to capture a bonanza of grant monies to examine the new technique and its impacts—are major Western oil and gas companies that now control only about 10 percent of the world's reserves.[48] They see carbon sequestration as a perfect business opportunity, and any large-scale commitment to carbon capture and storage would require their expertise in drilling, reservoir management, and gas and liquid transportation. Naturally, oil service companies (such as Baker, Halliburton, and Schlumberger), companies handling liquefied air products (Liquid Air, Linde Gas, Praxair), and large investment houses would also benefit. All of these await the windfall of sequestration spending mandated by new government regulation that may never come—

and that, if it does come, will not be able to prevent the doubling of prein-dustrial atmospheric CO_2 concentrations. Because other sequestration options—mineral carbonation, CO_2 capture from the air, and assorted activities aimed at increased carbon storage in plants or soils—have no greater chances of rapid and significant success, it is virtually inevitable (barring, of course, a prolonged global economic collapse or natural mega-catastrophe) that the atmospheric concentrations of CO_2 will rise above 450 ppm, and that the planet will experience more than just a very modest warming.

In the past I have often disagreed with what I thought were exaggerated and strident conclusions of Greenpeace reports—but in this instance I find the organization's five-point summary of problems with carbon capture and sequestration restrained and accurate.[49] CCS cannot provide enough storage in time to avoid further substantial increase in emissions; it will be a major consumer of energy, erasing half a century of efficiency gains in electricity generation; there will always be concerns regarding the safety of long-term storage and possibilities of leaks; it will be an expensive undertaking; and it will carry significant liability risks.

Even so, the recent embrace of CCS may yet amount to an unstoppable tide. The IPCC has not issued any revealing quantifications on the minimum energy needs compatible with a high quality of life so that efforts could be made to reduce usage to this minimum, but it has produced a special report on CO_2 capture and storage.[50] The U.S. establishment—government, industry, and the academy—is not pushing for a reduction in the country's vastly excessive per-capita energy use, which is twice as high as that of the richest European Union (EU) countries or of Japan, but it promotes "an aggressive goal" that would see "widespread deployment of CCS" beginning in eight to ten years.[51]

I see this kind of carbon capture and storage as an inferior solution, and, as I argued in the beginning, we should do our utmost to get the rankings of solutions right because of the enormous investments and environmental impacts that are at stake.

Carbon sequestration on a scale sufficient to affect the earth's climate—a geoengineering endeavor involving capture, compression, pipeline transport, and underground storage of at least 10 billion tons of CO_2 every year—would be a task of an unprecedented magnitude that is now considered by many to be not only an acceptable component of international efforts to limit the increase of atmospheric CO_2 concentrations, but perhaps one of the most effective means of doing so. Chances are high that this extraordinary promise will fail to deliver.

6

Liquid Fuels from Plants

There is nothing new about liquid biofuels for transportation.[1] Henry Ford, for example, was a great promoter of ethanol, and his famous Model T could run on gasoline, ethanol, or a mixture of the two fuels, much like today's touted Brazilian flex vehicles. Increased fuel demand during World War I boosted U.S. ethanol production, but the introduction of leaded gasoline and the increasingly inexpensive refining of crude oil left hardly any economic room for crop-derived ethanol.[2] This situation began to change only after OPEC's two rounds of large oil price increases in 1973–74 and 1979–81. Brazil's was the first serious entry into this new energy industry. The country's sugar cane–based National Alcohol Program (Proálcool) began in 1975, and eventually more than half of Brazilian cars were using anhydrous ethanol.

By contrast, U.S. activities remained slow moving and marginal in their overall impact. Commercial production of fuel ethanol began in 1980, and it took nearly fifteen years to surpass 5 billion liters; accelerated expansion began only in 2002; more than 15 billion liters were shipped in 2005, 25 billion liters in 2007 (see figure 6-1) and 35 billion liters (9 billion gallons) in 2008. Much higher targets lie ahead: In June 2007 the Senate passed an energy bill that mandated no less than 36 billion gallons of ethanol by the year 2022, a very impressive seventeenfold expansion of ethanol output in two decades. But even if gasoline consumption continued to grow no faster than it has since the year 2000, it would reach about 180 billion gallons by 2022.

Because the energy content of ethanol is only 65 percent that of gasoline, 36 billion gallons of ethanol would be equal to no more than about 13 percent of the likely 2022 gasoline demand. Clearly, this is not the way to emasculate OPEC. And even with the 2022 gasoline demand cut to half the expected level thanks to new (and, for America, rather draconian) fuel

FIGURE 6-1

U.S. ETHANOL PRODUCTION, 1980–2007

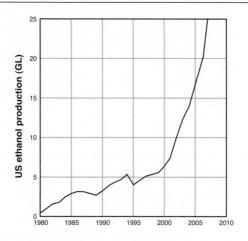

SOURCE: Plotted from data in Renewable Fuels Association (RFA, 2008).

efficiency requirements, and with the ethanol production mandate fully met, the share would rise to no more than a quarter of the total—again, no way to achieve the ever-elusive energy independence. But ethanol advocates ignore such likelihoods as they point out the fuel's environmental and security benefits.

Ethanol has been portrayed in highly positive terms: as the world's best choice for cutting dependence on crude oil and lowering the carbon intensity of the global economy;[3] as a perfect green energy solution ("Sugar beet, corn, wheat. Our recipe for renewable fuel," claims an advertisement from BP, the British global energy company); as an income stabilizer for grain farmers; as a fuel source that could get much better ("How much ethanol can a company extract from corn? . . . There's no telling how much mileage our efforts may yield down the road," boasts leading chemical company BASF, as though photosynthetic efficiency were infinitely elastic).

Europe, where nearly half of all passenger cars have diesel engines, is more smitten by biodiesel. This fuel can be extracted from a wide variety of oil-bearing crops, including such common temperate plants as rapeseed, sunflowers, and soybeans. In the tropics, it is oil palm and jatropha that many enthusiasts see as the means of liberating poor countries from their dependence on imported oil needed to fuel trucks, irrigation pumps, and small electricity generators. Such visions ignore the fact that the oil palm plantations have been a major cause of tropical deforestation, and that we know little about jatropha's domestication and long-term agronomic requirements, ignorance of which has already led to the abandonment of unrealistic plans for mass-scale jatropha cultivation.[4] Biofuel enthusiasts envision making use even of waste fats, among other suggestions for unusual—even bizarre—biofuel sources. These include kelp in California's waters and kudzu vines covering abandoned land in the U.S. Southeast, the harvesting of either of which would be no small challenge, and poppy seeds from Afghanistan—one need only look at the size of those seeds and check the yield of oil per unit of cultivated area to see how viable that proposal is! My assessment of liquid biofuel's prospects will concentrate first on the ethanol made from corn (by fermentation of carbohydrates that make up about 75 percent of the harvested grain) or sugar cane (by fermentation of sucrose expressed from cane stalks) that is to substitute for automotive gasoline.

Liquid Fuels for Transportation

In 2005, the worldwide demand for liquid transportation fuels was equivalent to about 2 billion metric tons of crude oil.[5] This demand was dominated by automotive gasoline (much more important than aviation gasoline for propeller-powered aircraft), by diesel fuel used for land transport (cars, trucks, off-road vehicles, railways) and marine transport, and by kerosene for jet-powered flight. Even if today's best commercially established and most productive biofuel alternative—Brazilian ethanol from sugar cane produced with a power density of 0.45 W/m^2—could be replicated throughout the tropics—an optimistic assumption—the land needed to produce transportation ethanol would add up to about 600 million hectares. That is more than all the land now cultivated in all tropical regions, and the equivalent of nearly 40 percent of the world's entire cultivated area.

Cultivation of lower-yielding subtropical and temperate phytomass could not approach sugar cane productivity, and hence the real land needs for replacing liquid transportation fuels by biofuels would be much larger—too large to allow for simultaneous harvests of the fuel and food needed for the eight and a half to nine billion people who will inhabit the planet by 2050.

Corn-Based Ethanol

Because of America's extraordinarily high gasoline consumption (equal in 2000 to nearly 80 percent of Japan's *total* energy use)[6] and the inherently low power density of ethanol production (only about 0.25 W/m^2 of cultivated land) corn-derived ethanol can never supply anything more than a relatively small part of the overall demand for fuel in the United States. If America's entire corn harvest—just over 280 million tons in 2005—were converted to ethanol at the best conversion ratio of 0.4 L/kg of grain, the country would produce fuel equivalent to 13 percent of total gasoline consumption.

Conversely, if all of America's gasoline demand were to be covered by corn-derived ethanol (produced at 0.25 W/m^2), the crop would have to be grown on some 220 million hectares of arable land, or on an area roughly 20 percent larger than the country's total arable land. Of course, a combination of higher crop yields, higher ethanol yields, and better average car performance should gradually lower this enormous land requirement—but even that may make little difference, as it is by no means certain that such gains would offset the rising demand. And land claims of corn-based ethanol would be much worse outside the United States—global corn yield averages just over 50 percent of the U.S. mean.

The disparities between the maximum theoretical potential of ethanol production and the actual volume needed to displace gasoline are obviously so large that corn-derived ethanol could become the dominant liquid fuel for U.S. transportation only if the current demand were cut by an order of magnitude. These disparities further mean that corn-derived ethanol can provide only a relatively small share of the overall need, even with significant future improvements in feedstock yields and conversion efficiencies, and even disregarding the many negative environmental impacts imposed by the entire production system.

Promoters of ethanol have been trying to minimize or trivialize these impacts while exaggerating the fuel's potential for easing America's oil import burden. But we should note that the country's recent ethanol drive is only partially a product of renewed concerns about high crude oil prices and Middle Eastern instability; its powerful agribusiness, with its enormous and effective Washington lobby, has strongly influenced the process. Three large agribusiness companies—Archer Daniels Midland (ADM), VeraSun, and Cargill—produce nearly 30 percent of U.S. ethanol[7] and have been doing so with massive federal subsidies, in yet another example of large private corporations eagerly scooping up generous public handouts.

Detailed examinations of both the overt and indirect instances of government support for ethanol and biodiesel identified annual subsidies worth $5.5 billion–$7.3 billion in 2006 in the United States (or as much as $1.39/gallon) and almost $5 billion in the EU; the studies also found that these subsidies continue to grow rapidly, both in scope and scale.[8] Although many other technical advances have received generous subsidies, the downsides of ethanol production are numerous, even setting the subsidies aside. They range from low net energy return on the entire process to a significant aggravation of human interference in the global nitrogen cycle.

Net Energy Analysis of Corn-Based Ethanol. The foremost consideration for any energy source is its net energy return (the ratio of the energy in the final product to the energy required to produce the commodity), and on that score, corn-based ethanol does not fare well. Pimentel's study, which accounts not only for the direct energy costs of corn cultivation but also for the energy costs of field machinery and irrigation, concludes that the ratio of energy contained in ethanol to energy used in corn production and fermentation is just 0.77, which is a significant energy loss.[9] Shapouri and others end up with a minimal gain of 1.06.[10] The highest positive energy returns, 1.56–1.67, are from studies that have given energy credits for byproducts of corn grain fermentation, mainly distillers' grain and corn gluten meal used for animal feeding.[11]

All of these studies have their shortcomings,[12] and hence it is impossible to come up with a single precise value, depending on the assumptions used. But there is no doubt that, at best, the whole exercise has only a marginal net energy gain, and corn-based ethanol would be a bad choice even if its net energy returns were unambiguously positive. The reason: Net energy studies entirely ignore the large-scale environmental degradation that would result from intensified and expanded corn cultivation and grain processing.[13]

Environmental Degradation. Corn is by far America's most common row crop, and before the closure of its canopy the soil is exposed to potentially heavy water erosion. Not surprisingly, corn cultivation is already the single largest source of the country's agricultural soil loss. High yields of corn also demand high applications of nitrogen fertilizers—on average more than 150 kg/hectare, and exceeding 200 kg/hectare in the Corn Belt.[14] But because typical uptake efficiency is less than 40 percent,[15] the crop is responsible for most of the nitrogen leached from the Mississippi basin to the Gulf of Mexico, resulting in the eutrophication of coastal waters and a larger dead zone in the gulf.[16] Corn irrigation is already the single largest user of underground water in the basin, and expansion of the corn-growing area into drier western fringes, or further intensification of corn production, would create additional demand for the mining of the already receding Ogallala aquifer.[17] Large volumes of wastewater from distilleries—ten to thirteen times the volume of the produced ethanol—also result in high biological oxygen demand and increase the energy burden of the fuel's production.

Grain corn cultivated overwhelmingly for animal feed has been traditionally rotated with soybeans, a combination that avoids massive monoculture and, at least in most cases, reduces the need for nitrogenous fertilizers. This beneficial rotation would disappear where corn was grown for ethanol, as contracts made with ethanol producers would call for a steady supply of grain for their operation and would result in more extensive corn monoculture across much of the United States. A policy promoting corn-based ethanol would also inevitably lead to the planting of corn on sloping or arid land that had previously been set aside for conservation. Once these ecosystemic realities are considered, it becomes obvious that ethanol production based on intensive corn cultivation is not a renewable activity, but an unsustainably extractive and environmentally detrimental enterprise.

Taken together, our consideration of its effect on the global nitrogen cycle and analysis of its net energy return make clear that massive corn-based ethanol production would not provide an overall economic, social, or environmental benefit. Interestingly, a comparative study of ethanol-powered transportation and vehicles energized by electricity produced by burning the same amount of biomass found that the latter choice yields greater reduction of greenhouse gas emissions.[18] And more intensive farming producing higher crop yields for increased ethanol conversion requires more nitrogen fertilizers and generates higher emissions of nitrous oxide, a much more powerful greenhouse gas than CO_2.

The net result is that biofuel production has already aggravated, rather than eased, greenhouse gas emissions.[19] Moreover, substituting a truly renewable biofuel (that is, one whose production would not need any fossil fuels) for gasoline would be even more demanding. All of the cited net energy calculations include a great deal of external, nonrenewable energy, such as coal, natural gas, and coal- and nuclear-derived electricity, which is used to produce the key inputs—fertilizers, pesticides, field machinery—that go into corn farming and to fuel the conversion process. Power densities of a largely renewable operation, in which the field machinery is fueled with ethanol that is fermented and distilled with heat derived by the combustion of crop residues, would drop the power density of the entire process to less than 0.1 W/m^2.

Sugar Cane–Based Ethanol

Sugar cane is obviously a much better choice as a feedstock for ethanol fermentation. This tropical grass photosynthesizes year round; its average global yield is now about 65 t/hectare (of which about 12 percent, or 7 t/hectare, is sucrose); and properly selected cultivars (thanks to endophytic nitrogen-fixing bacteria in stems and leaves) do not need any nitrogen fertilizer or only minimal supplementation. In addition, the ethanol production does not require any external fuel, as it can be entirely energized by the combustion of bagasse, the fibrous residue left once the juice is expressed from the cane stalks. All of this makes the cultivation of sugar cane and the subsequent ethanol fermentation a clearly energy-rewarding enterprise.

According to Macedo, Leal, and da Silva, typical cultivation and fermentation practices in the state of São Paulo, Brazil, have an energy return of 8.3, and the best operations can have rates just in excess of 10. Depending on which U.S. study is used for comparison, this is at least five and perhaps as much as ten times higher than for the corn-derived U.S. ethanol.[20] But a later study finds that Macedo and colleagues' research underestimates the energy cost of sugar cane cultivation and that, calculated correctly, the energy generated by the country's ethanol is only about 3.7 times that used to produce and distribute it.[21] In any case, in a rational global economy, there would be no ethanol production from temperate grains, and affluent nations would import ethanol from the tropics. Instead, the United States has high tariffs to discourage the Brazilian imports and thus subsidizes the inefficient domestic production even more heavily.

At the same time, any exports of ethanol from the tropics must be viewed with a great deal of caution. They are desirable only if suitable strains of sugar cane are grown on land that is not needed for food production—and few countries besides Brazil have such land available. Brazilian ethanol, moreover, may not be such an energy bargain.[22] Even setting the matter of net energy returns aside, sugar cane's potential to produce significant volumes of ethanol shows limits that are not much different from those of corn. The crop's global average yield is now about 65 t/hectare, and even if it were converted with the high average efficiency of 82 L/ton that now prevails in Brazil,[23] it would yield less than 5,500 L/hectare. The total area planted to sugar cane in tropical and subtropical countries was about 19 million hectares in 2005, and if all that were devoted to ethanol production, the annual fuel yield would be equivalent to less than 6 percent of the world's 2005 gasoline consumption. To cover the entire demand, sugar cane would have to be planted on some 320 million hectares—that is, on 20 percent of the world's arable land. But as high-yielding sugar cane can be grown only in the tropics, it would mean that about 60 percent of the total area now under cultivation in that region would have to be devoted to cane for ethanol.

Nor are sugar cultivation and harvest, particularly as practiced now, and sugar conversion environmentally benign. Preharvest burning of sugar cane removes some 80 percent of the plant's tops and leaves, which make up about 25 percent of the cane's phytomass; and while this burning makes hand harvesting safer and faster, it is a major source of air pollution.[24] New equipment, requiring higher energy inputs, would have to be developed for mechanical harvesting of green cane. And while the endophytic cane bacteria can provide as much as 190 kg N/hectare (and hence good crops can be grown not just for decades but even for centuries without any nitrogen applications), erosion losses will eventually result in substantial yield declines unless the nutritional difference is offset by fertilizer applications.[25]

Impacts of Ethanol Production

Future expansion of ethanol crops, whether corn or cane, will certainly fall far short of the published scenarios, but the concern is that in the process of pursuing some unrealistically high supply shares—say, by seeking to replace 10–20 percent of all gasoline with crop-derived ethanol—many countries will cause serious environmental, economic, and social dislocations. At least, the United States and Brazil have the luxury of abundant farmland, but only Brazil has the conditions that can support a relatively large and profitable biofuel industry. Low yields and the recurrence of drought in major grain-producing regions limit the potential in Australia and Canada, the other two major land-rich food exporters, and land scarcity eliminates the world's three most populous low-income nations—China, India, and Indonesia—as entries into large crop-based biofuel cultivation.

Rising commodity prices have shown how even relatively low-volume ethanol industries in land-rich countries can have major impacts, leading some economists to write about how biofuels could starve the poor.[26] In 2006, as the United States was diverting about 20 percent of its corn to ethanol, corn prices had risen to $4.20/bushel in the gulf ports by June, nearly 60 percent compared to the mean level of $2.69/bushel in 2005; and they were still close to $4/bushel by the fall of 2007.[27] The effect has been felt by the country's meat producers buying feed corn, and even more by importing nations that do not produce any biofuels. And a study by the Congressional Budget Office found that, between April 2007 and April 2009, ethanol production contributed as much as 15 percent to the increased cost of American food.[28]

These realities, and the use of food crops to produce fuel for export in some impoverished countries, led Jean Ziegler, the UN Special Rapporteur on the Right to Food, to issue a blunt assessment in October 2007, and to call for a five-year moratorium on the conversion of corn, wheat, and sugar into fuels: "It is a crime against humanity to convert agriculturally productive soil into soil which produces foodstuffs that will be burned as biofuel."[29] At the same time, the competitiveness of biofuels remains questionable even when the environmental externalities of their production are ignored. When operating without subsidies, ethanol production in some OECD countries is competitive only with oil prices ranging between $65

FIGURE 6-2
A SEGMENT OF CELLULOSE

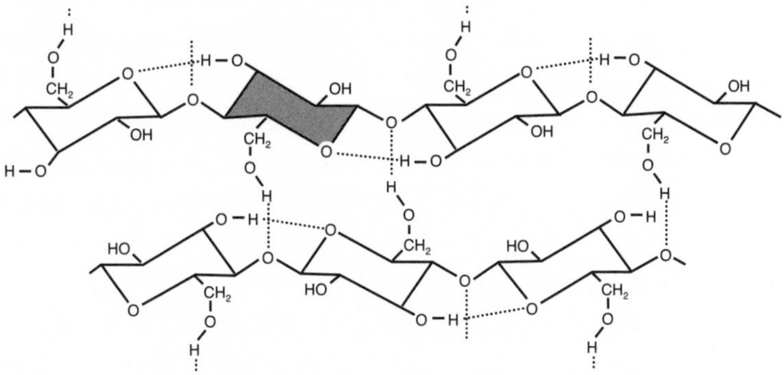

SOURCE: Smil 2000.

and $145/barrel.[30] Clearly, oil price fluctuations can greatly affect the profitability of biofuels.

Cellulosic Ethanol, "A Huge New Source of Energy"

The promoters of crop-derived ethanol, however, have what they see as a perfect answer to all these concerns: cellulosic ethanol—that is, alcohol fermented from sugars obtained by breaking down cellulose. The biosphere's most abundant macromolecule, cellulose is composed of about three thousand units of glucose (figure 6-2).

The choice of this substrate is compelling. Every crop—food, feed, or fiber, whether harvested for its seeds, tubers, leaves, or stalks—leaves behind a great deal of residual phytomass. Grains dominate crop farming in all affluent countries, and their modern cultivars have residue-to-grain ratios of roughly 1:1. The United States alone produces nearly 1 billion tons of crop residues annually, and the world's affluent economies harvest about 900 million tons of cereal grains every year and so have roughly the same amount of cereal straws and corn stover. The dominant polymer in this residue could be hydrolyzed to its constituent glucose molecules, and these could be fermented to ethanol, without making any additional claims on farmland.

Both public and private investment is elevating cellulosic ethanol from a bench-scale process to a mass commercial industry. The U.S. Department of Energy has invested in six cellulosic ethanol plants that should be completed by 2011, one based solely on corn stover, one on waste wood, and the rest on a mixture of agricultural wastes and waste wood.[31] And Silicon Valley entrepreneurs are the latest worshippers at the altar of cellulosic ethanol: Vinod Khosla, a cofounder of Sun Microsystems, has been a particularly eager promoter who claimed in 2006 that cellulosic ethanol would be cost competitive by 2009.[32]

But here again, energy dreams are unconnected to reality. Before a single commercial fast breeder was ever built, the proponents of the technique saw it dominating global electricity generation in just two decades; before a single Hypercar capable of 200 mpg ever left a factory, its promoters saw it conquering the automotive market in a matter of years. Likewise cellulosic ethanol: Now, before a single commercial facility producing cellulosic ethanol has entered routine and truly commercial operation (that is, paying its way), uncritical cheerleaders see that fuel as a huge new source of competitive energy.

Khosla sees cellulosic ethanol as a fuel that "is greener, cheaper, more secure than gasoline—and this shift won't cost the consumer, automakers or the government anything."[33] In the process, even the laws of thermodynamics get broken: we are promised something highly valuable for nothing. In reality, there are fundamental, and costly, challenges. The mass of crop residues that can be removed from fields without serious long-term environmental consequences is limited; the residues have inherently very low power densities, much lower than the crops; they are often difficult to collect and expensive to transport; and, once gathered, their structural cellulose and lignin are not easy to break down to produce fermentable sugars.

Crop residues are not valueless wastes that have been waiting for biofuel enthusiasts to turn them to ethanol; they are an extremely valuable resource that provides a number of indispensable and irreplaceable agroecosystemic services.[34] In poor populous countries, crop residues still provide a great deal of fuel, animal feed, and fibrous material, and their assiduous recycling should be one of the key universal pillars of responsible agronomic management. Recycled crop residues return to the soil the three macronutrients of nitrogen, phosphorus, and potassium, as well as many micronutrients. They replenish soil organic matter—all healthy soils are assemblages of mineral constituents and of a multitude of living and dead microbes and invertebrates—and they retain moisture through their sponge-like action. They also prevent both wind and water erosion.

Proper agronomic practices thus see crop residues as a highly valuable, renewable resource whose indiscriminate repeated removal would have prompt and dramatic environmental and agronomic consequences. Therefore, only a carefully determined share of these residues should be removed from fields, and, in many instances, that small amount would not be worth collecting.

A closer look at America's most abundant crop residue illustrates the difficulties in setting the allowable removal limits, and no smaller challenges in harvesting and transporting the feedstock. U.S. corn cultivation, the world's largest and most productive, yields annually about 200 million tons of corn stover.[35] About half of stover is in the form of stalks, about a fifth is leaves, and the rest is almost equally divided between cob and husk. Less than 5 percent of U.S. stover is regularly harvested as cattle feed; the rest is plowed in or left on the surface to capture moisture and to prevent soil erosion. Stover availability will obviously fluctuate with annual yield; but even with a fixed harvest, it would not be easy to pinpoint the total that could be repeatedly removed from fields without ill effect. Standard assumptions use a stover-to-grain ratio of 1:1, but Pordesimo, Edens, and Sokhansanj suggest that a more conservative ratio of 0.8:1 (fresh weight) is a more realistic mean at the grain harvest moisture of 18–31 percent.[36] The general assumption is that when fully complying with the best management practices, 3–4 tons of stover could be harvested per hectare on a sustainable basis.

This conclusion may be overly optimistic; for example, Blanco-Canqui and others find that stover harvesting at rates above 1.25 t/hectare changes the hydraulic properties of soil and has a pronounced effect on earthworm activity.[37] They conclude that site-specific information, so far very limited, would be needed to establish permissible levels of stover harvest. Different stover-to-grain ratios and different assumptions about average moisture content and recycling requirements have resulted in estimates of annual U.S. stover harvests as low as 64 million and as high as 153 million tons of dry matter.[38] A fairly conservative approach is to assume that with conventional tilling, about 35 percent of stover could be removed from fields without any adverse consequences and that the rate rises to about 70 percent for no-till farming, for a weighted national mean of about 40 percent. Thus, about 80 million tons (dry weight) of stover could be removed annually, a theoretical equivalent of no more than 3 percent of today's U.S. gasoline consumption. Clearly, biofuel from stover is not a way to reduce significantly the country's dependence on imported oil.

And, in any case, field availability does not equal ethanol plant input. Mass industrial processes require predictable and uniform inputs, but crop residues fit neither requirement. Their yields will fluctuate with crop yields, and so will their composition. Thomas analyzed more than 1,100 stover samples of more than a hundred corn hybrids in ten states and found the range of total structural carbohydrates (45–69 percent of dry matter) to be large enough to affect the minimum ethanol selling price by as much as 25 percent.[39] Harvesting of stover is nearly always a challenge.[40] Chopping it up to maximize its field drying is done easily, but drying is obviously an unpredictable function of local weather. Days or weeks may elapse before stover dries to baling moisture, and during rainy spells it may never happen. Farmers would have to gather phytomass that is more than a third water and thus pay more for transporting this excess water to processing plants.

Another problem in harvesting is that a shredder may collect less than two-thirds of stover on the field. Collection would be made easier if shredding could be routinely combined with windrowing—that is, with raking the stover into rows. But to facilitate field drying of a fairly moist phytomass (40–50 percent water), the stover is spread as widely as possible. A round baler collects only about half the shredded, and preferably drier, stover, so that the overall harvesting efficiency is only about 30 percent.

Whatever the harvesting rate, chopped stover loaded in trucks has a very low density, just around 70 kg/m^3, and it needs to be compacted to make its collection logistically more affordable. If deliveries within an 80-km radius cost \$30–\$35/ton of dry matter, then transportation of freshly harvested stover to conversion plants would add to the inefficiency of the process, as the trucks would be hauling 30 percent water. Moreover, because the conversion of stover to ethanol would obviously require large-scale, months-long storage, the resource availability would be further diminished; dry matter losses of baled stover would range between 10 and 25 percent.[41]

For that reason, scenarios of high-residue harvest are predicated on the development of new techniques to harvest large shares (70–75 percent) of available residues, but such high rates would be tolerable in most places only with no-till farming. Perlack and colleagues conclude that high-residue harvests would be possible only if *all* U.S. cropland were cultivated without tilling and if new techniques enabled farmers to harvest 75 percent of all field residues. They call no-till cultivation "the most environmentally friendly production system."[42] That, too, is a revelation: How could no-till monocultures (replacing the current plow agriculture dominated by corn-soybean rotations) planted on all cultivated land and requiring repeated applications of herbicides be so friendly?

And after all these travails, there is still the critical challenge of converting cellulose into glucose. Enzymatic hydrolysis is the preferred way to release glucose molecules from cellulose and ferment them to produce ethanol. The cellulases required for enzymatic hydrolysis are naturally possessed only by a relatively small number of organisms, most notably by fungi and bacteria, and the hydrolysis is not easily accomplished; if it were, then cellulose, the most abundant structural constituent of terrestrial phytomass, would be subject to rapid degradation and decay. Cellulases were discovered only during the 1940s, and serious work at scaling up their batch production and lowering their cost began only during the 1990s. The chemistry and engineering required by large-scale enzymatic conversion of cellulose are complex, and, as a result, no commercial cellulosic ethanol plant was in operation at the end of 2007.[43] Even if the six demonstration plants funded by the U.S. Department of Energy enter operation by 2011, and even if they perform as expected, challenges of scaling-up lie ahead. The combined capacity of these plants would be equivalent to just 0.1 percent of transportation fuel used in the United States in 2005.[44]

The theoretical yield of ethanol from cellulosic crop residue is now about twice as large as the actual rate. Narrowing the difference would require finding better ways to remove lignin, whose presence occludes the target polysaccharides and inhibits their enzymatic hydrolysis, as well as finding a yeast that can efficiently ferment both hexoses and pentoses—especially xylose from the hydrolysis of hemicelluloses—and that would also tolerate higher ethanol levels before requiring expensive separation.[45] All these challenges run head on against the basic structural and functional properties of plants. These properties were selected for by evolution, and it would be naïve to underestimate the difficulties involved in their modification. In addition, new ethanol-producing facilities are fairly costly, and at this point we have no clear idea about future cost trends.

Global economic recession has further slowed down the efforts to commercialize cellulosic ethanol, and the best estimate in 2009 for its still basically experimental production capacity in the United States was less than 5 percent of the capacity previously projected by the U.S. Environmental Protection Agency for 2010.[46] Responsibly optimistic American experts think that biofuel production from cellulosic phytomass will be realized only in the next ten to fifteen years.[47]

The conclusion is thus fairly clear: Even if we had batches of astonishingly potent and inexpensive cellulases and could operate new plants with very low costs, it would take several decades before the production of cellulosic ethanol could be scaled up to commercial level, and finding sufficiently potent enzymes would not be the only decisive factor. Any large-scale extension of high-yielding cellulosic crops faces many uncertainties, and hence no confident forecasts are possible. For example, the highly touted American switchgrass does not have a long history of monocultural cultivation, and some surprisingly large uncertainties pertain to its nutrient requirements and best agronomical practices.[48] Moreover, switchgrass and other rhizomatous grasses considered for biofuel production have many traits that make them a potentially highly invasive species.[49] A detailed integrated energy, environmental, and financial analysis further found that ethanol produced from switchgrass must rely heavily on nonrenewable energies (mainly for fertilizer, herbicide, and machinery); hence, it is not a good substitute for petroleum products.[50]

As for biodiesel, its production is much less developed than ethanol fermentation; it accounts for just 15 percent of the global biofuel total, with Europe being the dominant producer. I cannot resist deconstructing one specific proposal for biodiesel production: making it from spent coffee grounds. Its authors calculated that the process of oil extraction and transesterification would produce nearly 1.3 billion liters (340 million gallons) of diesel fuel—that is, if all of the world's spent coffee grounds were assiduously collected and processed in economically sized facilities.[51] What they forgot to mention is that the global production of diesel fuel now exceeds

800 billion liters, and hence their scheme would produce an equivalent of less than 0.2 percent of today's global diesel fuel consumption—hardly a green energy source to save the planet.

Sadly, biofuel promoters have also begun to advocate the use of animal lipids. Apparently, their understanding of energy conversions has yet to reach the undergraduate level, as they utterly ignore the immense inefficiency of turning animal feed into stored fat. When I first read about using salmon oil as biofuel,[52] I thought there could be no more bizarre suggestion. There are two critical points. First, wild salmon stocks have been severely reduced by overfishing, and hence any additional salmon catch for fuel oil would spell the final death toll for this precariously surviving species. Second, farmed salmon requires roughly 3.1–3.9 units of fishmeal and fish oil to produce a unit of edible tissue,[53] and this feed must be obtained by catching massive amounts of such wild species as sardines, anchovies, and shrimp. Consequently, one could hardly think of a better way to completely destroy several once superabundant marine species, or produce a hugely negative energy outcome, than producing salmon oil fuel.

Regrettably, these kinds of delusions are publicly funded, some handsomely so. Among the most absurd ideas is an International Energy Agency (IEA) program that evaluates "the risk of using animal tallow derived from specified risk materials, dead stock, and downer animals as feedstock for the production of biodiesel."[54] Just imagine: relying on biofuel from "risk material" or, to put it plainly, mad cows. Do the authors expect a panzootic of bovine spongiform encephalopathy? Another proposal would "link a biodiesel plant with the cosmetic surgeons."[55] This is the Earthrace project, founded by New Zealander Peter Bethune, which aims to set a new round-the-world powerboat speed record in a boat powered by biodiesel fuel partly manufactured from human fat. According to Bethune, "In Auckland we produce about 330 pounds of fat per week from liposuction, which would make about 40 gallons of fuel."[56]

Biofuels, An Inappropriate Solution

Finally, the most obvious consideration that militates against liquid biofuels as an energy source for modern transportation is their utter lack of system appropriateness. Even the biofuels produced with the highest possible efficiency and the least environmental impact should not be poured into vehicles whose performance has more in common with pre-1950 machines than might be expected in the early twenty-first century. The clearest indicator of this indefensible state is that the United States has utterly failed to improve its average motor vehicle fuel efficiency after it doubled the average performance between 1973 and 1985.[57]

If that rate of improvement had been continued after 1985—and this would have presented no insurmountable technical challenges—new cars would have averaged about 50 mpg by 2010, and the mean for the entire car fleet could be well above 40 mpg by 2015, more than halving the current U.S. need for automotive fuel and sending oil prices into a tailspin. European carmakers do not have an exemplary record, either: By 2000, the curb weight of their average compact car was 50 percent higher than in 1970.[58]

We must conclude that biofuels produced by existing conversions or by methods that are about to enter commercial applications are incapable of replacing refined oil products needed by today's road vehicles, ships, and airplanes, and they cannot, and should not, meet a large share of the global demand that will be made by more efficient transportation fleets during the next few decades. We must further conclude that a rationalization of the entire transportation system—its prime movers, machines, and organization, as well as expectations about its future—should precede even a carefully considered, and hence inherently limited, production of biofuels. Using complicated, energy-intensive, environmentally disruptive, and actually nonrenewable processes to produce liquid fuels for oversized, highly inefficient machines—which are operated all too often for dubious reasons—adds up to compounded irrationalities.

Although many second looks have tarnished the image of crop-based biofuels during the past five years, new unrealistic claims are now being put forward, above all ones regarding the potential for producing biofuels from algae. In this "gold rush for algae," their authors extrapolate bench-scale experiments under perfectly controlled growing conditions to massive outdoor ponds, some advocates even suggesting yields that are thermodynamically impossible.[59] This only confirms that the contest between energy myths and realities never ends.

My brief analysis has only skimmed across some major problems of crop-based biofuel production; those readers who want to look at perhaps the most comprehensive appraisal of biofuels should consult Giampietro and Mayumi's recent work.[60] After reading it they will understand why the authors used "biofuel delusion" to title their examination of the fallacy of large-scale production of liquid fuels from plants. The conclusion is thus clear: More important than the fact that liquid biofuels *cannot* displace refined oil products in transportation is that they *should not*.

7

Electricity from Wind

In a paper published in the *Journal of Geophysical Research* in May 2005, Christina L. Archer and Mark Z. Jacobson of Stanford University quantified the world's wind power potential.[1] Using wind speed measurements from about 7,500 surface stations and 500 balloon-launch stations on five continents, they interpolated wind speeds at 80 meters above the ground—the altitude equal to that of the hub of a large modern 1.5 MW turbine with a rotor diameter of 77 meters.

The study found that at that height, wind speeds average 8.6 m/s over the ocean and 4.5 m/s over land, and that about 13 percent of all locations with wind measurements have class 3 winds, that is winds with speeds in excess of 6.4–7 m/s when measured 50 meters above ground. These winds are strong enough for low-cost, large-scale commercial electricity generation. After assuming that the statistics generated from all analyzed stations were representative of the global wind distribution, and after applying an unrealistically high load (capacity) factor of 48 percent, they concluded that the global wind power that could be harnessed at the height of 80 meters in locations with mean annual wind speeds equal to or in excess of 6.9 m/s could generate 630 PWh, corresponding to the power of 72 TW.

The study confirmed the well-known regional differences in the distribution of wind power—Atlantic Europe (see figure 7-1) and the Great Plains of North America have the best conditions for wind-generated electricity—and its finding of a large aggregate potential strengthened the arguments in favor of a massive expansion of wind power. In 2008, global consumption of all forms of primary energy—fossil fuels, hydroelectricity, and nuclear electricity—amounted to 11.3 billion tons of oil equivalent, or roughly a fifth of the total that could theoretically be produced by wind, and the worldwide generation of electricity added up to 17.5 PWh, or to less than 3 percent of wind's global potential.

FIGURE 7-1

WIND SPEEDS IN COASTAL ATLANTIC EUROPE

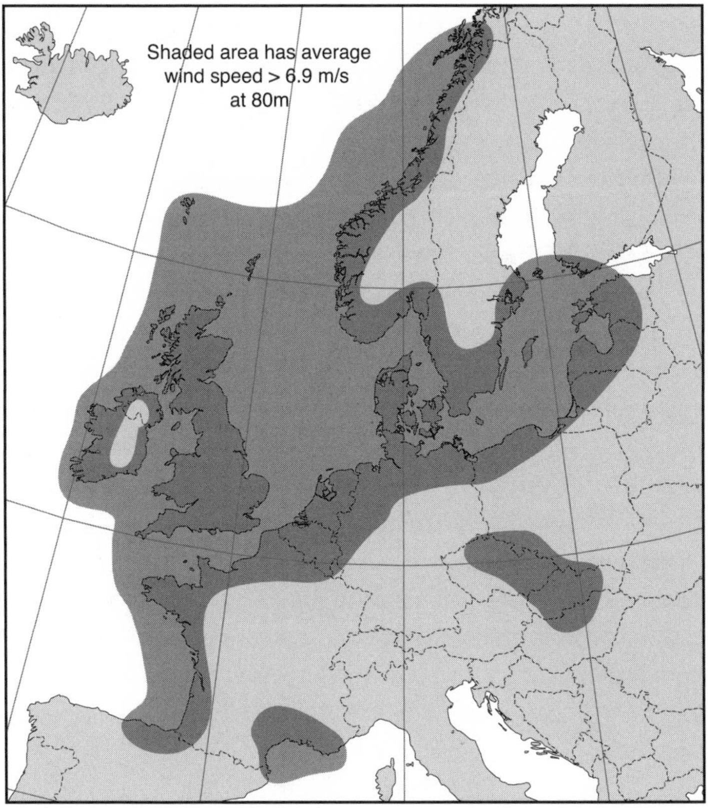

Shaded area has average
wind speed > 6.9 m/s
at 80m

SOURCE: Simplified from a map of European wind speeds in Archer and Jacobson (2005).

Predictably, the claim that "converting as little as 20 percent of potential wind energy to electricity could satisfy the entirety of the world's energy demands,"[2] which appeared in the American Geophysical Union press release announcing the publication of the Archer-Jacobson paper, was widely repeated and used as a very convincing illustration of wind's role as a possible global energy savior. Even before that global assessment was released, environmentalist Lester Brown had concluded from an earlier appraisal of U.S. wind potential[3] that "wind power can meet not only all U.S. electricity needs, but all U.S. energy needs." Brown had even called it "likely" that "wind power could satisfy not only world electricity needs but perhaps even total energy needs."[4]

Of course, these astonishing claims do not offer any particulars; it would be especially interesting to see how wind would replace coke in smelting the civilization's dominant metal. But renewable energy enthusiasts do not dwell on such details. They are convinced that wind's grand turn has definitely arrived, and that its conversions will change the global energy outlook.

And four years after Archer and Jacobson's study, they got an apparently even stronger argument for their claims. An appraisal based on a simulation of global winds 100 meters above the ground and using 2.5 MW turbines ended up with an even higher wind potential. After excluding areas covered by permanent ice and snow, forests, water, and built-up land, it put the potential wind generation for all continents at 680 PWh (power of 78 TW) when assuming an average 20 percent capacity factor.[5] In national terms, the study concluded that the U.S. potential is nearly twenty times larger than the current electricity generation, and that the difference for China is about sixtyfold, even when excluding all offshore sites. But a closer look at these enormous totals makes it clear that the world will not abandon all other forms of electricity generation anytime soon, and that wind-generated electricity will not dominate the global demand ten, or twenty, or twenty-five years from now.

Evolution of Wind Power

Indisputably, on a civilizational time scale of thousands of years, wind is an infinitely renewable resource, but until recently its conversions were done by small, inefficient machines. Mechanical power produced by windmills before the modern era was globally insignificant, although its local and regional contributions were notable,[6] particularly during the early decades of industrialization. Cheap fossil fuels and hydroelectricity ended any further diffusion of this energy capture, and it was not until the 1990s that conversions of wind's kinetic energy into electricity finally began taking off, particularly in Europe.

The first years of the twenty-first century have seen the elevation of wind from one of many renewable energy flows to a prime candidate for electricity supply in the post–fossil fuel world, a trend marked by some admirable advances in the design of large wind turbines and by an impressively rapid growth in wind-powered electricity generation. The unit capacity of commonly installed wind turbines has risen from less than 50 kW in the early 1980s to more than 1 MW two decades later and to more than 2 MW in Denmark, a leading pioneer of modern wind turbine design. By 2008, the largest prototype machine (German ENERCON's E-126) had a rotor diameter of 126 meters and a rated capacity of 6 MW, but the largest machine advertised on the company's website in 2009 was a 2 MW ENERCON E-82 model.[7]

Wind turbines with ratings of 1–2 MW are now concentrated in large groupings. The largest wind farms (such as Horse Hollow in Texas, which rates 735 MW) have installed capacities equal to that of midsize coal-fired electrical stations. And the world's largest planned offshore project, the London Array, nearly 20 km from the Kent-Essex coast and scheduled for completion in 2011, will have 340 turbines with an aggregate capacity of 1 GW, equal to that of a large thermal station and to 1 percent of the United Kingdom's total electricity needs. Obviously, there is nothing small, simple, or decentralized about these machines and facilities, and hence this fastest increasing mode of modern electricity generation cannot be classed among the Lovinsian soft-energy conversions thought to be coming to rescue the sinning civilization reliant on fossil fuels.

The global total of installed wind-generating capacity rose from 4.8 GW in 1995 to 17.4 GW in 2000 and 59.1 GW in 2005; at the end of 2008, it had reached 120.791 GW. Europe continued to lead in wind power with nearly 66 GW—about 55 percent of the global total—with Germany and Spain heading the aggregate national totals and Denmark still far ahead of any other country in installed per-capita capacity. U.S. wind-power capacity rose from just 10 MW in 1981 to 25.1 GW by the end of 2008—a 2,500-fold increase in twenty-seven years.[8]

Estimated Potential of Wind Power

Before deconstructing claims about potential wind generation, and before putting the recent turbine and capacity growth rates into appropriate context, I must first point out that the publication of Archer and Jacobson's aggregate estimate of 72 TW of global wind potential[9] did not come as a revelation to those familiar with the fundamental metrics of global atmospheric circulation. The authors claimed that the goal of their study was to quantify the world's wind power potential for the first time, but that quantification had been done before, albeit by deducing the total from first principles rather than building it up from thousands of wind measurements.

The theoretical derivation of wind's aggregate potential begins with the surprisingly small share of insolation—that is, solar radiation reaching earth—needed to drive global atmospheric circulation. Peixoto and Oort estimated that energy transferred to wind and dissipated as friction is less than 900 TW; Lorenz put the share higher, at about 2 percent of total solar radiation.[10] These totals set absolute theoretical limits on the availability of wind energy.

The strongest winds, in the powerful and shifting jet stream, are the most difficult to harness. They blow at altitudes around 11 km above the surface, and in the Northern Hemisphere their latitudinal location shifts with seasons between 30° and 70°N. Most people would consider the harnessing of these winds as pie in the sky, but the president of Sky WindPower Corporation believes they can be used to produce electricity for a mere cent/kWh at average U.S. locations, a fraction of the cost of coal-fired generation.[11] To achieve that result, 60 kW helicopter-like flying electric generators tethered by aluminum lines would have to be massively deployed.

There are other proposals for airborne generation. The Canadian company Magenn Power promotes giant floating wind turbines whose horizontally turning rotors would be supported by helium-filled balloons connected to the ground with tethers more than 300 meters long. In 2007 the company's (failed) plan was to offer 10 kW units (priced at $3–$5/W) by 2008; in 2009, it promised a 100 kW unit in 2010–11.[12] Wind power enthusiasts point out that there is enough energy in high-altitude winds to power civilization a hundred times over.[13] Well, there is enough energy in the solar radiation reaching the earth's ground to power civilization ten thousand times over—but neither source will energize our civilization anytime soon. The overall resource magnitude is no indicator of how rapidly or how easily it can be tapped.

Staying closer to the ground (conceptually and actually), we can see that only the winds moving in the lowest few hundred meters above the surface are practical candidates for interception. On the global scale, about 35 percent of wind energy, or no more than roughly 1.2 PW, or nearly 2.5 W/m², is dissipated within 1 km of the surface. Gustavson, justifying his choice "as a compromise between caution and imprudence in the face of inadequate knowledge," opted for 10 percent of this near-surface dissipation as an upper limit of practical wind energy utilization that could not be exceeded without adversely altering global atmospheric circulation and changing climate.[14] This reasoning results in a global wind potential of about 120 TW. This means—especially as Archer and Jacobson concluded that their 72 TW total was likely on the low side[15]—that a very similar estimate of wind's potential power has been available for nearly three decades.

Key Constraints on Wind Power

Now for the deconstruction of the Archer-Jacobson total. I will deal only with the fundamental considerations of resource and reserve dichotomy, power density, intermittency, and capacity integration, and not with many secondary matters that have undoubtedly been restraining the diffusion of wind turbines and that may, in the future, combine to impose significant limits on the ultimate development of wind-powered electricity generation. These considerations include the aesthetic impact of massive wind farms,[16] wind turbine noise, and the threat to birds and bats. While the gearbox and generator noise in turbines has been nearly eliminated, and the blade noise is usually acceptable with adequate exclusion zones, many people will have strong objections to visual pollution, and many conservationists are unhappy about inevitable bird kills[17]—though so far, tall buildings, windows, high-voltage (HV) wires, and cats far surpass wind turbines as causes of bird mortality.[18]

I will also not consider the limits imposed by generation costs. Published calculations try to demonstrate that the total costs of wind energy have for years been less than those of coal energy even in the United States,[19] while other analyses have disputed that conclusion.[20] Similarly, in Europe, Awerbuch's risk-adjusted estimates of generating costs show that, over their lifetimes, the wind turbines being installed today will produce electricity at a lower cost than natural gas–fueled generation,[21] while others point to the need for continuing substantial subsidies. Claims and counterclaims get complicated by including or omitting credits for preventing carbon emissions or eliminating other pollution impacts. Estimates are also complicated by the verdict on subsidies: Are they a laudable policy to help a superior technique in its early market penetration stages, or are they an unwarranted market distortion?

That matter is important in judging the costs of wind-powered generation, because its history has been closely tied with considerable subsidies. They have helped launch both the European and U.S. wind industries, and an interesting case can be made in favor of their further sensible expansion: Rather than subsidizing European and North American farmers to produce excessive harvests of crops that distort the global agricultural market and burden the environment with unnecessary impacts—particularly the leaching of fertilizers—it would be more rational to subsidize the mass-scale construction and operation of wind farms on agricultural land. That would create a new form of farming income while still leaving most of the land available for cropping or return to natural vegetation, and it would also help lower dependence on energy imports.

Resource and Reserve. I will leave all these secondary considerations aside and start the deconstruction by outlining the gap between the aggregate wind resource and the practically usable share of that resource—that is, the wind equivalent of fuel energy reserves. Power estimates of 70–120 TW of wind in class 3 and higher correspond to the resource category for mineral deposits (for example, to oil in place) of which only a part—typically no more than a third without enhanced recovery methods—could be extracted. In the case of wind, the limitation on what is practically usable is even greater. In some hydrocarbon fields, enhanced oil recovery using water flooding or gas lift can extract well over 50 percent, even 70 percent, of all oil in place; but it is quite obvious that we will not be able to harness 50 or 70 percent of the overall wind potential.

The analogy with hydropower potential offers perhaps the most useful corrective. If the potential energy of the global runoff were to be used with 100 percent efficiency, the gross theoretical capability of the world's rivers would be about 12 TW; this aggregate would be conceptually analogical to the 72 TW of wind potential estimated by Archer and Jacobson.[22] Competing water uses, the unsuitability of many sites, seasonal fluctuations of flow, and the impossibility of converting water's kinetic energy with perfect efficiency at full capacity mean that the exploitable capability—that is, the share of the theoretical potential that can be tapped with existing techniques—will be a small fraction of the theoretical availability. Aggregation of detailed national assessments put the globally exploitable capacity of flowing water at roughly 14 percent of the theoretical total.[23] But not everything that is technically feasible is economically acceptable, and the sites that meet the latter criterion add up globally to only about 8 percent of the theoretical potential and only about three times the total that was actually generated by 2009.

One important consideration in any estimate of mass-scale wind-driven generation is that we do not know the maximum share of global atmospheric circulation that could be converted into electricity without changing the earth's climate. The most obvious physical restriction on wind exploitation arises from the fact that giant turbine farms cannot be erected in many suitably windy places. Their construction and operation are either outright impossible (in urban settings) or economically questionable (in rugged or remote terrain), or they are simply highly undesirable (in protected areas such as natural parks and scenic shorelines). Moreover, the construction of giant wind farms will continue to encounter local objections, even in many otherwise suitable locations.

Detailed mapping can eliminate all locations unsuitable because of physical obstacles, but there is no objective, definite process for defining the practical potential within or near protected areas and, of course, no way to anticipate which proposed site may see speedy construction of a wind farm and which will be forced to cancel a project because of resolute opposition. For these reasons, it is very difficult to quantify the actual magnitude of wind reserves—but even if we stipulate that wind power available within 100 meters above the surface is more than sufficient to satisfy today's global electricity demand, it is highly unlikely that even half that capacity total will be ever realized, as enormous space claims will constrain future expansion.

Power Density. Much as large hydrostations that cut the middle and lower courses of rivers must be supported by large reservoirs (thus resulting in the very low power density of hydrogeneration), large wind farms have extraordinarily large space demands. Even in windy regions (power class 4, 7–7.5 m/s at 50 meters above ground) such as the Dakotas, northern Texas, western Oklahoma, and coastal Oregon, where wind strikes the rotating blades with power density averaging 450 W/m^2, the necessary spacing of wind turbines (at least five, and as much as ten, rotor diameters apart, depending on the location, to reduce excessive wake interference) creates much lower power densities per unit of land. For example, a large 3 MW Vestas machine with a rotor diameter of 112 meters spaced six diameters apart will have peak power density of 6.6 W/m^2, but even if an average load factor were fairly high (at 30 percent), its annual rate would be reduced to only about 2 W/m^2.

Actual peak rates are highly site specific. The most densely packed wind farms rate in excess of 10 W/m^2 of land, and more spread-out sites typically range between 5 and 7 W/m^2—though they can rate as low as 1 W/m^2— while the best offshore sites with the highest winds have power densities greater than 15 W/m^2.[24] A realistic approximation for typical year-round loads (not peak power densities!) of today's large-scale wind farms would thus be around 2 W/m^2. This density would be valid for future regional-sized wind farms only when their extraction of wind energy would not significantly alter large-scale winds—but both mesoscale and global-scale models of atmospheric circulation indicate that the very large-scale extraction of wind (requiring installed capacities on a TW scale needed to supply at least a quarter of today's demand) reduces wind speeds and consequently lowers the average power density of wind-driven generation to around 1 W/m^2 on scales larger than about 100 km.[25]

Global electricity production reached roughly 18 PWh in 2007, and this output required nearly 4 TW of installed capacity, which prorates to an average load factor (the number of hours in a year when a turbine actually generates electricity) of just over 50 percent—the rate between the higher load factor for thermal plants (coal-fired > 70 percent; nuclear > 85 percent) and the lower rate for hydrostations (<40 percent). Wind-driven generation has a much lower load factor. Although the range of 30–35 percent has been frequently assumed in recent literature, the most complete examination of the actual record for the EU, the world's largest concentration of wind power, shows that during the five years between 2003 and 2007 the capacity factor amounted to less than 21 percent.[26] This means that the cost of wind power is two-thirds higher and the reduction of CO_2 emissions is 40 percent lower than was previously assumed.

It also means that in large-scale calculations of wind-powered electricity generation, we should not assume average load factors higher than 25 percent. Supplying half of today's electricity—that is, about 9 PWh—by wind would thus require about 4.1 TW of wind turbines; with 2 W/m², they would claim about 2 million km², or an area roughly four times the size of France or larger than Mexico. With average power density of just 1 W/m², the required area would rise to more than 4 million km², roughly an equivalent of half of Brazil or the combined area of Sudan (Africa's largest country) and Iran.

These calculations indicate that deriving substantial shares of the world's electricity from wind would have large-scale spatial impacts. Obviously, only a small portion of those areas would be occupied by turbine towers and transforming stations, so that crop planting and animal grazing could take place close to a tower's foundations. But even when assuming a large average turbine size of 2–3 MW, the access roads (which are required to carry heavy loads, as the total weight of foundations, tower, and turbine is more than 300 tons per unit) needed to build roughly 2 million turbines and new transmission lines to conduct their electricity would make a vastly larger land claim than the footprint of the towers; and a considerable energy demand would be created by keeping these roads, often in steep terrain, protected against erosion and open during inclement weather for servicing access.

Given the extraordinarily high U.S. electricity demand, then, any chances of a wind-powered U.S. economy seem remote. The U.S. energy infrastructure, including the right of way for all high-voltage transmission lines, now occupies up to about 25,000 km², or 0.25 percent of the country's area, roughly equal to the size of Vermont.[27] And the country's entire impervious surface area of paved and built-up surface reached about 113,000 km² by the year 2000.[28] In contrast, relying on large wind turbines to supply all U.S. electricity demand (about 4 PWh) would require installing about 1.8 TW of new generating capacity, which (even when assuming an average of 2 W/m²) would require about 900,000 km² of land—nearly a tenth of the country's land, or roughly the area of Texas and Kansas combined.

Wind Intermittency. These calculations, realistic in terms of space claims, do not address the fundamental matter of wind power's intermittency. With modern wind turbines, this is not, as it used to be with many old windmills,

FIGURE 7-2

POWER CURVE OF 3-MW TURBINE V90 WITH WIND SPEED CATEGORIES

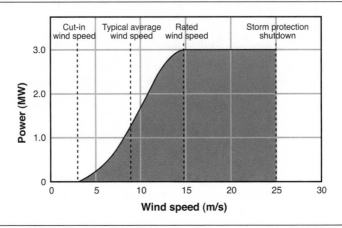

SOURCE: Vestas 2007.

a matter of all or nothing. Modern turbines begin generating as soon as wind speeds reach 3–4 m/s, or roughly 10–14 km/h; and because wind power goes up with the cube of wind speeds, every subsequent doubling of speed results in an eightfold increase in power. Rated power is reached at 12–14 m/s (43–50 km/h), and it is sustained until the wind speed reaches 25 m/s (90 km/h), when the machine is shut down to protect the blades and the tower in stormy weather (see figure 7-2). As a result, a large modern turbine situated in a windy site will generate 70–85 percent of the time, but its output will vary from a small fraction of its rated capacity to its maximum power.

Many studies have demonstrated that these variations cause no unmanageable problems, even in an isolated electricity generating system, as long as the total power installed in wind turbines is no more than about 10 percent of the system's overall output. The British utility company National Grid Transco concluded that "the expected intermittency of wind does not pose . . . a major problem for stability and we are confident that this can be adequately managed."[29] In larger, well-interconnected systems, any load variability caused by wind's intermittency will be lessened by siting wind turbines in many locations sufficiently far apart to make a concurrent becalming of a large share of operating machines highly unlikely, or by relying on diversified generation, using coal-fired, nuclear, hydro, and wind generation.

These conditions work clearly to Europe's advantage. The continent's extensive long-distance high-voltage interconnections and its diversified generation mean that as much as 20 percent, and eventually perhaps even 30 percent, of its total capacity could be contributed by wind turbines without resorting to any excessive buildup of reserve capacity (now most often gas turbines that could be rapidly activated). A study by the German Energy Agency concluded that the planned 14 percent share of wind generation in the country's total electricity consumption by the year 2015 will not require construction of additional power stations to balance the increasing contribution of wind-generated electricity. It further concluded that security of supply for the system can be well maintained, and that only a minor expansion of the national grid (850 km of new extra-high-voltage lines and upgrading of another 400 km) will be required.[30]

But this good news has its definite limits, a fact ignored by uncritical wind promoters. The European Wind Energy Association's website deals with "the intermittency myth" by claiming that "there is little overall impact if the wind stops blowing somewhere—it is always blowing somewhere else"[31] True, but that "somewhere else" may be hundreds or thousands of miles away with no high-voltage transmission lines in between. A new worldwide system where wind would be the single largest source of electricity would require such vast intra- and intercontinental extensions of HV transmission lines to create sufficiently dense and powerful interconnections to deal with wind's intermittency that both its cost and its land claims would be forbidding.

Østergaard's study of geographic aggregation and wind power output variance makes this clear. Drawing on the Danish experience, he finds, predictably, that demand and wind variations in different areas help even out fluctuations and reduce imbalances in systems with high reliance on wind power, and that exploiting these variations allows for reductions in reserve capacity in other modes of electricity generation. But, no less predictably, he also finds limits to what can be done: The average requirement for the reserve thermal capacity may drop, "but the same is not generally the case with the maximum required condensing mode capacity. . . . There will simply be times with wind production in neither of the interconnected areas."[32]

Conversely, there are times when there is an excess of wind-generated electricity, and when dumping it on interconnected neighbors will depress the price. In the Danish case these situations have led to a bizarre new development, as Nord Pool, the Nordic electricity-trading system, introduced a negative spot price for electricity (€200/MWh) starting in October 2009.[33] This amounted to penalizing the coal-, hydro-, or nuclear-based generators for Denmark's excess wind electricity on the market.

Many region-specific climate peculiarities also limit the maximum share of wind power in a system's portfolio. Archer and Jacobson found that North America is particularly well suited for wind-powered generation: It has the largest number of stations in class ≥3 of all the inhabited continents, and highly windy sites are not concentrated in a single region but are found along the Atlantic coast from Newfoundland to North Carolina, around the Great Lakes, in a broad midcontinental swath from Manitoba to Texas, and in the West along the coasts of California, Washington, British Columbia, and Alaska.[34]

The North American continent also, however, has a relatively high frequency of both prolonged calms and excessively strong winds, and the Southeast is affected by both. Prolonged calms are created in the region during summer and early fall by the semi-stationary high-pressure cell centered west of Bermuda. This Bermuda high is associated with calm or very slow winds, limited mixed-layer formation (hence conducive to air pollution buildup), and high temperatures. Local wind generation is thus at a minimum while electricity demand for air conditioning is at its annual maximum, and because it would be impossible to rely on wind power during this period, the region would have to import large blocks of electricity from the Great Lakes region or from the Midwest—but this arrangement would require a number of additional long-distance high-voltage lines.

Moreover, during summer and early fall, global circulation brings frequent hurricanes that can affect the coastal and nearby inland regions extending from Texas to Nova Scotia. These would require repeated shutdown of all wind-generating facilities for a number of consecutive days and would repeatedly expose all turbines and their towers to serious risk of damage and possible prolonged repairs. A perfect example of these consequences was the severe damage sustained by many offshore oil drilling rigs in the Gulf of Mexico during Hurricane Katrina in 2005. Of course, one can argue that turbines should simply not be sited in these risky regions and that the needed power should come from the continent's interior, where the machines would not be exposed to hurricanes, though they would remain vulnerable to frequent tornadoes.

Capacity Integration. After consideration of the siting of wind turbines, we now turn to the matter of adequate interconnections, which in theory looks fairly promising. A study by the National Renewable Energy Laboratory found that the United States has 175 GW of potential wind capacity located within five miles of existing lines carrying up to 230 kV, 284 GW within ten miles of such lines, and 401 GW within twenty miles of such lines.[35] But what matters more than distance to the nearest transmission line is that line's capacity, and in this respect it is obvious that the situation in the United States is much inferior to that in Europe. Europe has strong and essentially continent-wide north–south as well as east–west connections, while the United States does not have a comparably capable national network: high-voltage connections from the heart of the continent, where the wind potential is highest, to either coast are minimal or nonexistent.

Consequently, the Dakotas could not become a major supplier to California or the Northeast without massive infrastructural additions. Jacobson and Masters argue that with an average cost of $310,000/km (an unrealistically low mean; see the next section), the construction of 10,000 km of new HV lines would cost only $3.1 billion, or less than 1 percent of the cost of 225,000 new turbines, and that HV direct current lines would be even cheaper.[36] As with any entirely conceptual megaproject, these estimates are highly questionable; moreover, such an expansion is not very likely, given that the existing grid (aging, overloaded, and vulnerable) is overdue for extensive, and very expensive, upgrading,[37] and that securing rights of way may be a greater challenge than arranging the needed financing.

That is why such proposals as Cavallo's scheme for a 2 GW wind farm in Kansas connected by a 2,000-km-long high-voltage link with California to replace California's Diablo Canyon 2.2 GW nuclear power plants remain unrealistic. This project, assumed by Cavallo to be online by 2010, was to have a high-capacity factor of 60 percent, which large-scale compressed-air storage was to boost to an incredibly high rate of 70–95 percent.[38] As always with such megaprojects, theoretical calculations are one thing, engineering realities another; in 2010 there are no such schemes in the United States, or anywhere else.

Similar considerations apply in the Chinese case. Recent reports see China becoming the world's largest wind energy market by the year 2020, when at least 20 GW (and even 40 GW) are to be installed, with plans for

400 GW by 2050.[39] But China's wind generation potential is highly concentrated in two regions: along the southeast coast, particularly in Fujian Province, and in the northwestern interior, mainly in Nei Monggol. China's most populous province, however, is the landlocked Sichuan, with most of its population living in a large, intermountain basin hidden by fogs on most satellite images. These fogs are due to one of the world's highest frequency of calms. In Sichuan's capital, Chengdu, the average annual frequency of calms is 42 percent (compared to 20 percent in Beijing and 10 percent in Shanghai), and the winter mean is close to 50 percent—obviously a very unfavorable condition for any large-scale, wind-powered electricity generation for the more than 100 million people in the province.[40] As in the U.S. case, any large-scale reliance on wind-generated electricity would be predicated on first putting in place long high-voltage, high-capacity transmission lines.

Realizing the Potential of Wind Power

How far and how fast wind-powered generating capacities will increase is uncertain. In 2007, wind turbines produced about 1.25 percent of the world's electricity, with the highest national shares in Denmark (about 21 percent), Spain (nearly 12 percent), Portugal (just over 9 percent), Ireland (over 8 percent), and Germany (7 percent); the U.S. share remained below 2 percent.[41] If the worldwide growth in wind turbine installations were to continue at the rate that has prevailed since 1995, then wind farms would surpass the 2006 capacity of all electricity generating plants (fossil fuel–fired, nuclear, and hydro) in fifty years. Clearly, that is not going to happen, as the high growth rates characteristic of early stages of growth, whether of organisms or new techniques, will moderate, and a logistic curve will form.

But while it is exceedingly improbable that the world of the 2050s will have nearly 4 TW of wind capacity capable of generating nearly 9 PWh of electricity (equal to roughly half of today's worldwide generation from all sources), it is impossible to specify the most likely asymptote—the ultimate capacity of wind power that will be captured by large wind turbines. Many factors—ranging from the cost of competing supplies and the extent of the eventual resurrection of the nuclear option, to concerns about the rate of global warming and the level and duration of subsidies—will determine the eventual outcome.

Some long-term goals are very ambitious. Pacala and Socolow estimate that using wind power as one of the substitution wedges needed to stabilize CO_2 emissions would require the installation of about 2 TW of peak wind capacity by 2054 to displace 700 GW of coal generation, or 4 TW of peak wind power to displace 1.4 TW of natural gas–fueled generation.[42] These aggregates would call for installing, respectively, 40 GW/year for fifty years, compared to the 2000–2008 annual addition rate of less than 13 peak GW. *Wind Force 12*, a report prepared by the European Wind Energy Association (EWEA) and Greenpeace, projects 3.04 TW of global wind capacity in 2040 and envisages that wind would generate about 22 percent of the world's electricity, with an average load factor of 30 percent.[43] In absolute terms, this is nearly half the total generated by all means in 2006. Similarly, a later report by Greenpeace and the European Renewable Energy Council calls for only a slightly lower global share by the year 2050.[44] But these aggregates should be questioned: To reach 3.04 TW by the year 2040, the wind-generating capacity increase during the four decades between 2000, when the total installed capacity was less than 20 GW, and 2040 would have to amount to just over 3 TW, requiring a sustained annual installation pace nearly six times higher than the 2000–2008 mean.

Another way to understand this high installation rate is to compare it with the worldwide nuclear and hydroelectric capacity increases between 1960 and 2000, the first of which amounted to 350 GW and the other to about 600 GW. But even when considering the easily replicable modularity of wind turbines and their much faster rates of installation, we are still left with a taxing and relentless pace of introduction, and we cannot be sure if it can be achieved and maintained; installing some 750 GW per decade is an unprecedented challenge. Moreover, as we have no long-term experience with the operation of massed large turbines, we cannot be certain about their average life spans (will all of them work steadily for twenty-five or thirty years?) or their lifetime needs for maintenance and replacement (will all major components last, even in harsh offshore environments, for at least twenty years?).

Whatever the eventual and as yet unknowable aggregate of wind capacity increments may be, technical realities and the imperatives of large-scale grid management make the following matters clear: An isolated national grid could draw most of its electricity from wind-powered generation only if it had massive, multi-GW storage, a condition that does not exist in any country. In the case of nations with abundant and relatively persistent wind flows, whose small to midsize grids also have high-capacity interconnections to systems supplied with thermal (both fossil and nuclear) power and hydro-electricity, like Denmark, it should be possible to raise the share of wind power to as much as 40 percent, exceptionally perhaps even 50 percent.

Regions with high-capacity, high-density, high-voltage grids already largely in place (such as northwestern Europe), or windy regions where such grids could be gradually constructed (America's Great Plains and the Canadian prairies), could eventually derive as much as 30 percent of their electricity from wind. Globally, we may aspire to about 15 percent by 2030 or 2040, while 30 percent is quite unrealistic, and 50 percent is simply impossible. The conclusion is clear: Conversion of wind's kinetic energy by large turbines can become an important contributor to the overall electricity supply, but, except for relatively small regions, it cannot become the single largest source, even less so the dominant mode of generation.

8

The Pace of Energy Transitions

Not all news concerning America's energy challenge was gloomy during the summer of 2008, though oil prices rose to nearly $150/barrel, raising the country's July crude oil import payments to nearly $42 billion—compared to $22 billion in July 2007—and creating even more anxiety about the country's dependence on foreign oil.

Craig Venter, a pioneer in the sequencing of the human genome, announced that the scientists at his institute had created the first synthetic bacterial genome,[1] another key step toward the completely synthetic bacterium-like organism that Venter's Synthetic Genomics aspires to design for the production of ethanol or hydrogen.[2] And T. Boone Pickens, one of America's most famous billionaires, began to promote his energy transition plan.[3]

Released in July 2008, the Pickens plan got a great deal of attention because of its promoter's background: An octogenarian oilman who had made a fortune in the Texas oilfields was advocating a retreat from oil and spending his own money to do it. Pickens advertised widely, appeared on many TV shows, testified before Congress, and then returned with follow-up TV advertisements seeking public support for his proposal. The greatest appeal of the Pickens plan to reduce America's dependence on foreign oil was its cascading simplicity.

First, Pickens wanted to dot the Great Plains ("the Saudi Arabia of wind power") with enough wind turbines to replace all the electricity currently produced by burning natural gas. Second, he wanted to use the freed-up natural gas to run efficient and clean natural gas vehicles. Third, he believed that this substitution would create a massive new domestic aerospace-like industry that would offer well-paying jobs producing giant turbines and auxiliary equipment and bring economic revival to the depopulating Great Plains. Fourth, he further believed that this substitution would reduce the huge outflow of wealth to oil-producing nations, as under his plan the United States would cut its imports of oil by more than one-third. And, Pickens claimed, he was committed to spending his own money to get the process going, by building the country's largest (4 GW) wind farm in West Texas.

Also released in July 2008, just as oil prices peaked at $147/barrel, was Al Gore's call for a rapid, radical replacement of America's entire thermal electricity generation industry by green alternatives.[4] Gore expressed no doubts either about the plan's incredibly short time frame or about its economic feasibility: "Today I challenge our nation to commit to producing 100 percent of our electricity from renewable energy and truly clean carbon-free sources within 10 years. This goal is achievable, affordable and transformative. . . . To those who say 10 years is not enough time, I respectfully ask them to reconsider what the world's scientists are telling us about the risks we face if we don't act in 10 years." Gore saw only two options for "those who, for whatever reason, refuse to do their part": They "must either be persuaded to join the effort or asked to step aside."[5]

As I will show, the proposals by Gore and Pickens have much in common with similar recent promises, forecasts, and visions of the imminent and profound difference to be made by new energy conversions. All ignore one of the most important realities ruling the behavior of complex energy systems: the inherently slow pace of energy transitions.

Present Realities

I have already deconstructed or alluded to a number of promises similar to those of Gore and Pickens. By the year 2000, coal-based generation of electricity was to be a relic of the past, with all demand supplied by nuclear fission and with the superefficient breeder reactors already taking over; by the year 2000, between 30 and 50 percent of America's energy use was to come from renewable flows; by the year 2000, the world was to derive half its energy from natural gas. And a decade ago, the promoters of fuel cell cars were telling us that by now such vehicles would be on the road in large numbers, well on their way to displacing ancient and inefficient internal combustion engines.

These are the realities: Coal-fired power plants produce almost 50 percent of U.S. electricity and nuclear stations about 20 percent. All the

FIGURE 8-1

PREDICTED AND ACTUAL TRENDS IN U.S. DEPENDENCE
ON FOREIGN CRUDE OIL

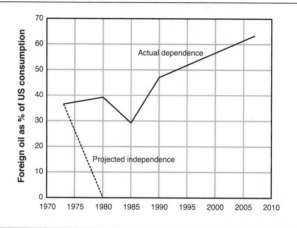

SOURCE: Data points calculated from consumption and import statistics in British Petroleum (BP, 2008).

nuclear stations are first-generation, water-cooled fission reactors; not a single commercial breeder reactor is operating anywhere in the world. In 2008 the United States derived less than 2.5 percent of its energy from new renewables—that is, from corn-based ethanol, wind, or photovoltaic solar or geothermal power.[6] Natural gas provided 24 percent of the world's commercial energy, not the 50 percent share predicted in the early 1980s, which means that it is still less important than coal, which in 2008 supplied 29 percent of the world's commercial primary energy.[7] And there are no fuel cell cars to be bought anywhere.

A revealing illustration of the blunders committed by ignoring the gradual nature of energy transitions is offered by another famous energy plan for America, announced by President Richard M. Nixon in November 1973 and reiterated in his State of the Union address in January 1974: "Let this be our national goal: At the end of this decade, in the year 1980, the United States will not be dependent on any other country for the energy we need to provide our jobs, to heat our homes, and to keep our transportation moving."[8] In 1973, the country was importing just over a third of its crude oil; in 2008 it bought nearly 70 percent (figure 8-1). Gore's repowering plan follows in the unrealistic tradition of Nixon and later of President Jimmy Carter, who, famously fond of wearing an energy-conserving cardigan, said in July 1979: "Beginning this moment, this nation will never use more foreign oil than we did in 1977," as he reset the energy independence date to 1990.[9]

Past Transitions

The point has been clearly made: All the forecasts, plans, and anticipations cited above have failed so miserably because their authors and promoters thought the transitions they hoped to implement would proceed unlike all previous energy transitions, and that their progress could be accelerated in an unprecedented manner. Today's advocates and promoters obviously think the same. Could they be right?

To answer this question, we need a simple definition first: An energy transition encompasses the time that elapses between the introduction of a new primary energy source (coal, oil, nuclear electricity, wind captured by large turbines) and its rise to claiming a substantial share of the overall market. This "substantial share" is necessarily arbitrary, though I would argue for at least 15 percent, or roughly every seventh unit of total supply, because the equivalents of shares lower than 10 percent can usually be achieved by demand adjustments and do not require new technical solutions; 20 percent or 25 percent would obviously be a more decisive contribution. Obviously, for a new entrant to become the single largest contributor, it must have a share higher than 33 percent among three supply components, or higher than 25 percent among four. For it to be an absolute leader, it must contribute more than 50 percent of the energy supply. While there are no such fuels or electricity sources on the global scale, many examples exist on national scales.

Some fairly good historical data make it possible to identify the tipping points of the first great energy transition, from the millennia-long reliance on biomass fuels like wood, charcoal, or crop residues to coal or, later, a mixture of coal and crude oil. In the United States, it was only in the early 1880s that the energy content of coal (and some oil) consumption surpassed the energy content of fuel wood. The best available historical reconstruction points to the late 1890s, when half the world's energy came for the first time from the combustion of fossil fuels and all but a small fraction of that from

FIGURE 8-2
GLOBAL SHARES OF COMMERCIAL PRIMARY ENERGIES, 1900–2008

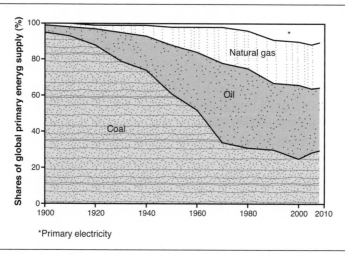

*Primary electricity

SOURCE: Based on Smil (2008b) and British Petroleum (BP, 2008).

coal. In Russia, that point came no earlier than the late 1920s, and in China sometime during the 1960s; and in a number of African countries, traditional biomass fuels still continue to dominate the overall energy supply.[10]

For fossil fuels on the global scale, coal receded from about 95 percent of the total energy supply in 1900 to about 60 percent by 1950; it was surpassed by oil only in 1965, and it had declined to less than 24 percent by 2000. But even then its importance continued to rise in absolute terms, and in 2001 it began to regain some of its relative importance. Today, coal, which provided nearly 29 percent of primary energy in 2008, is more important in relative terms than it was at the time of the first energy "crisis" in 1973, when it provided about 27 percent; and in absolute terms it now supplies twice as much energy as it did in 1973. The world (thanks largely to China and India, as well as to massive Australian and Indonesian exports) has been returning to coal rather than leaving it behind (see figure 8-2).[11]

Crude oil had become the largest contributor to the world's primary energy supply by 1965, and although its share reached as much as 48 percent by 1973, its relative importance then began to decline, and in 2008 it contributed less than 37 percent. Moreover, during the twentieth century coal contributed more energy than any other fuel, edging oil by about 5 percent. The common perception of a nineteenth century dominated by coal and a twentieth century by oil is wrong. In global terms, 1800–1900 was still a part of the millennia-long wooden era, and 1900–2000 was (albeit by a small margin) the coal century. And while many African and Asian countries use no coal, the fuel remains indispensable worldwide in many ways: It generates 40 percent of the world's electricity and 50 percent of the U.S. total, and it supplies nearly 80 percent of all energy in South Africa, the continent's most industrialized nation, 70 percent in China, and almost 60 percent in India.[12]

The pace of the global transition from coal to oil can be judged from the following spans: It took oil about fifty years from the beginning of its commercial production during the 1860s to capture 10 percent of the global primary energy market and then almost exactly thirty years to go from 10 percent to about 25 percent of the total. And it took natural gas no less than seventy years (1900–1970) to rise from 1 percent to 20 percent of the total. Since that time, natural gas has been the fuel with the highest increases in annual production, but by 2008 its share was, as already noted, only about half what had been expected in the 1970s, and at 24 percent it was below that of coal.[13]

As far as electricity is concerned, hydrogeneration began in the same year as Edison's coal-fired generation (1882). Just before World War I, water power produced about half the world's electricity; its subsequent fast and sustained expansion in absolute terms could not prevent a large decline in its relative contribution, which by 2008 was about 17 percent. Nuclear fission also ascended rapidly, reaching a 10 percent share of global electricity generation just twenty-seven years after the commissioning of the first nuclear power plant in 1956. Its further growth, however, largely stopped during the 1980s, and its share is now roughly the same as that of hydro power.[14]

Energy transitions involve not only new fuel sources but also the gradual diffusion of new prime movers—that is, devices that replace animal and human muscles by converting primary energies into mechanical power, which can then be used to rotate massive turbogenerators producing electricity, or to propel fleets of cars, ships, and airplanes. Transition times from established prime movers to new converters have been often remarkably long. Steam engines, whose large-scale commercial diffusion began in the

1770s with James Watt's improved design, remained important into the middle of the twentieth century. There is no more convincing example of their endurance than the case of the Liberty ships, the "ships that won the war," as they carried American materiel and troops to Europe and Asia between 1942 and 1945.

Rudolf Diesel began to develop his highly efficient internal combustion engine in 1892, and his prototype engine was ready by 1897. The first small ship engines were installed on river-going vessels in 1903, and the first ocean-going ship with diesel engines was launched in 1911. By 1939, a quarter of the world's merchant fleet was propelled by those engines, and virtually every new freighter had them—but 2,751 Liberty ships were still powered by large, triple-expansion oil-fired steam engines.[15] And steam locomotives disappeared from American railroads only in the late 1950s, while in China and India they were indispensable even during the 1980s.

The adoption of automotive diesel engines is another excellent proof of the slow pace of energy transitions. The gasoline-fueled internal combustion engine, the most important transportation prime mover of the modern world, was first deployed by Benz, Maybach, and Daimler during the mid-1880s, and it reached a remarkable maturity in a single generation after its introduction (Ford's Model T in 1908). But massive car ownership came to the United States only during the 1920s, and in Europe and Japan only during the 1960s, meaning that thirty to forty years in the U.S. case and seventy to eighty years in the European case elapsed between the engine's initial introduction and its decisive market conquest, with more than half of all families having a car. The first diesel-powered car (Mercedes-Benz 260D) was made in 1936, but it was only during the 1990s that diesels began to claim more than 15 percent of the new car market in major EU countries and only during this decade that they began to account for more than a third of all newly sold cars. Once again, roughly half a century had to elapse between the initial introduction and significant market penetration.[16]

Similarly, it took more than half a century for any internal combustion engine, either gasoline or diesel fueled, to displace agricultural draft animals in industrialized countries. The U.S. Department of Agriculture stopped counting draft animals only in 1963, and the substitution of engines for animals has yet to be completed in many low-income nations. Finally, when asked to name the world's most important continuously working prime mover, most people would not name the steam turbine. The machine was invented by Charles Parsons in 1884, and it remains fundamentally unchanged 125 years later. Gradual advances in metallurgy simply made it larger and more efficient, and these machines now generate more than 70 percent of the world's electricity in fossil-fueled and nuclear stations, with the rest coming from gas and water turbines and diesels.[17]

Why Energy Transitions Are Gradual

No common underlying process explains the gradual nature of energy transitions. In the case of primary energy supply, the time span needed for significant market penetration is mostly a function of financing, developing, and perfecting necessarily massive and expensive infrastructures. For example, the world oil industry handles about 30 billion barrels annually, or 4 billion tons, of liquids and gases. It extracts the fuel in more than a hundred countries, and its facilities range from self-propelled geophysical exploration rigs to sprawling refineries and include about 3,000 large tankers and more than 300,000 miles of pipelines.[18] Even if an immediate alternative were available, writing off this colossal infrastructure that took more than a century to build would amount to discarding an investment worth well over $5 trillion—and it is quite obvious that its energy output could not be replicated by any alternative in a decade or two.

In the case of prime movers, there is often inertial reliance on a machine that may be less efficient (steam engine, gasoline-fueled engine) than a newer machine but whose marketing and servicing are well established and whose performance quirks and weaknesses are well known; the concern is that rapid adoption of a superior converter may bring unexpected problems and setbacks. Predictability may, for a long time, outweigh a potentially superior performance, and the diffusion of new converters may be slowed down by complications associated with new machines. One such complication pertains to the high particulate emissions of early diesels; another arises from new supply-chain requirements—for example, sufficient refinery capacity to produce low-sulfur diesel fuel, or the availability of filling stations dispensing alternative liquids.

All energy transitions have one thing in common: They are prolonged affairs that take decades to accomplish, and the greater the scale of prevailing uses and conversions, the longer the substitutions will take. Although the second part of this statement seems to be a truism, it is ignored as often as the first part; otherwise, we would not have all those unrealized predicted milestones for electric or fuel cell cars or for clean coal or renewable conversions. These realities should be kept in mind when appraising potential rates of market penetration by nonconventional fossil fuels, by new biomass fuels, or by renewable modes of electricity generation.

The Repowering Challenge

None of the alternatives named has yet reached even 5 percent of its global market. Nonconventional oil, mainly from Alberta's oil sands, now supplies only about 3 percent of the world's crude oil and only about 1 percent of all primary energy.[19] Renewable conversions—mainly liquid biofuels from Brazil, the United States, and Europe, and wind-powered electricity generation in Europe and North America, with much smaller contributions from geothermal and photovoltaic electricity generation—now provide about 0.5 percent of the world's primary commercial energy.[20] The relevant U.S. production rates were virtually nothing for nonconventional crude oil and about 4 percent for crop-derived ethanol as a share of gasoline demand; less than 1.5 percent of all electricity comes from wind-powered generation and about 0.02 percent from solar conversions.[21]

But is not today's situation fundamentally different? Do we not possess incomparably more powerful technical means to effect faster energy transitions than we did a century or a half century ago? We do—but we also face an incomparably greater scale-up challenge. While the shares of new energies in the global or the U.S. market remain negligible, the absolute quantities needed to capture a significant portion of the total supply are huge because the scale of the coming global energy transition is of an unprecedented magnitude. By the late 1890s, when combustion of coal (and a bit of oil) surpassed the burning of wood, charcoal, and straw, each of the two resource categories supplied annually an equivalent of about half a billion tons of oil. If during the coming decades we sought to replace worldwide only 50 percent of all fossil fuels with renewable energies, we would have to displace fossil energies equivalent to about 4.5 billion tons of oil, a task equal to creating *de novo* an industry whose energy output would surpass that of the entire world oil industry that took more than a century to build.

If we are guided by Gore's specific goals, it is rather easy to quantify America's repowering challenge. In 2008, the country generated about 3.75 PWh in fossil-fueled and nuclear stations, the two nonrenewable forms of generation that Gore wants to have entirely replaced by renewable conversions. Installed capacity in these stations was about 870 GW, which means that their load factor was almost exactly 50 percent, and it took the country fifty-seven years to add this capacity.[22] In 2008, the wind and solar electricity generating industries contributed 1.2 percent of the total, and with installed capacity of about 25 GW, their load factor averaged just 24 percent.[23] Accordingly, even if all requisite new HV transmission interconnections were in place, slightly more than two units of generating capacity in wind and solar would be needed to replace a unit in coal, gas, oil, and nuclear—and the country would have to build about 1,740 GW of new wind and solar capacity *in a decade*, 1.75 times as much as it built during *the past fifty or more years*.

But that is not all. If achievable, such a feat would mean writing off in a decade the entire fossil-fueled and nuclear-generation industry, an enterprise whose power plants alone have replacement value of at least $1.5 trillion; and (assuming an average cost of about $1,500/kW) it would also mean spending at least $2.5 trillion to build the new capacity. Conceivably, the first feat can be achieved by some accounting sleight of hand; but where will deeply indebted and financially precarious America get $2.5 trillion to invest in this new generating infrastructure within a single decade? And because those new plants would have to be in areas not currently linked with HV transmission lines to major consumption centers (wind from the Great Plains to the east and west coasts, PV solar from the Southwest to the rest of the country), that "affordable" proposal would also require, as Gore himself admits, a massive rewiring of the United States.

Limited transmission capacity to move electricity eastward and westward from what is to be the new power center in the Southwest, Texas, and the Midwest is already delaying new wind projects even as wind generates less than 2 percent of all electricity. The United States now has about 212,000 miles of HV lines, and inadequacy of the country's poorly interconnected grids is a major bottleneck for a rapid development of wind and solar generation capacities, while the American Society of Civil Engineers estimates that an investment of $1.5 trillion would be needed by the year 2030 to improve the grid's reliability and connectivity.[24]

But the eventual cost is bound to escalate, given that the regulatory approval process alone is likely to take many years before new line construction can begin. In sum, it is nothing but a grand delusion to think that in ten years the United States can achieve wind and solar generation whose equivalent in thermal power plants took nearly sixty years, while incurring write-off and building costs on the order of $4 trillion, concurrently expanding its electricity grid by at least 25 percent and modernizing the rest—while also reducing regulatory approval of megaprojects from many years to mere months.

False Analogy

But Gore would argue that the plan is doable and affordable because "as the demand for renewable energy grows, the costs will continue to fall." He then goes on to give the key specific example:

> The price of the specialized silicon used to make solar cells was recently as high as $300 per kilogram. But the newest contracts have prices as low as $50 a kilogram. You know, the same thing happened with computer chips—also made out of silicon. The price paid for the same performance came down 50 percent every 18 months—year after year, and that's what's happened for 40 years in a row.[25]

Gore implies that, analogically, the costs of photovoltaic electricity generation could be halved every eighteen months for decades to come.

But the comparison is wrong, and the implication is impossible. To begin with, if the cost of photovoltaic cells were to decline by 50 percent every eighteen months for just ten years, their cost at the end of that period would be just about 1 percent of the starting value, and the modules, now retailing for nearly $5/W, would be selling before the year 2020 for just $.05/W; we would then be close to producing electricity too cheap to meter. And the comparison is functionally wrong, as well. Moore's law, the doubling of microprocessor

FIGURE 8-3

GRAPHIC PRESENTATION OF MOORE'S LAW

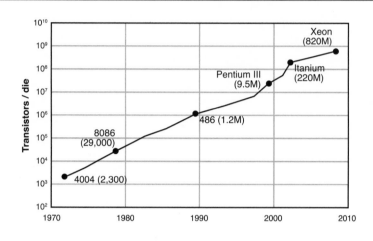

SOURCE: Based on data in Intel (2007, 2010).

performance every two years with ensuing price declines,[26] has worked primarily because of an ever-denser packing of transistors onto silicon wafers—from 2,250 transistors for Intel's first microchip in 1971 to 820 million transistors per die for its latest dual-core processors in 2007 (see figure 8-3)[27]—not because of cheaper crystalline silicon. After all, a blank silicon wafer is worth only about 2 percent of the total value of a finished microprocessor.

Undoubtedly, PV cells have been getting cheaper. Modules cost more than $20 per peak watt in 1980, about $10 by 1985, and around $5 a decade later; but the price was still close to $4.50 at the end of 2009.[28] Moreover, their performance, even from the perspective of the best rates in research settings, has not been improving by orders of magnitude. In 1980, the best thin-film cells were about 8 percent efficient, and by 1995 the efficiency had doubled to about 16 percent; but by 2010 was only about 20 percent, while the performance of the more expensive multijunction concentrating monocrystalline cells rose from about 30 percent in 1995 to about 40 percent by 2010 (see figure 8-4).[29]

Consequently, even the best conversion rates achieved in research settings have doubling periods of fifteen to twenty years, not fifteen to twenty

FIGURE 8-4

INCREASING CONVERSION EFFICIENCIES OF PHOTOVOLTAIC CELLS

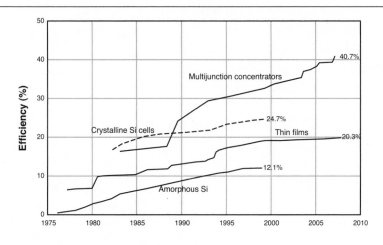

SOURCE: National Renewable Energy Laboratory. 2010. Best research-cell efficiencies. www.nrel.gov/pv/thin_film/docs/kaz_best_research_cells.ppt.

months, and inherent physical limits will make it extremely difficult, if not impossible, to ever achieve yet another doubling for multijunction and monocrystalline cells. Moreover, the PV industry now aims at reducing the price of solar modules from about $4.5/W by the end of 2009 to $1.5–$2/W within a decade, a rate of price improvement far more sluggish than that conforming to Moore's law. And the cells themselves are only part of the overall cost, which also includes their mounting in modules, batteries, inverters, and regulators (adding up to about 80 percent of the final cost) and installation (accounting for the rest). According to surveys by Solarbuzz, a company that researches and consults on solar energy, the price of PV electricity generated by a small (2 kW) residential system declined only 10 percent between the end of the year 2000 and the end of 2009, from nearly 40 c/kWh to just over 35 c/kWh. Similarly, even the electricity produced by the largest (500 kW) industrial systems was only 7 percent cheaper in late 2009 than in late 2000.[30]

The doubling of microprocessor performance every two years is an atypically rapid case of technical innovation that does not represent the norm of technical advances as far as new energy sources and prime movers are concerned. Inherent physical limits restrict efficiency gains to a doubling or, at most, a tripling of the current values for today's low-performance (thin-film and amorphous) PV cells during the next ten to twenty-five years, and, similarly, unit costs may be halved or quartered during similar periods of time.

Moreover, Gore's single-decade leap greatly underestimates the task of building new transmission links to carry electricity from the country's windiest states (North Dakota is at the top) and sunniest states (Arizona) to large cities on both coasts (see figure 8-5). He concludes that "the cost of this modern grid—$400 billion over 10 years—pales in comparison with the annual loss to American business of $120 billion due to the cascading failures that are endemic to our current balkanized and antiquated electricity lines."[31]

Characterizing the U.S. transmission grid as balkanized and antiquated is quite correct, and it is also true that the new HV underground cables insulated with cross-linked polyethylene (XLPE), which are increasingly being chosen for new HV transmission links, have become considerably cheaper.[32] But the scaling-up challenge would still be enormous. In 2008, the total worldwide length of these connections (both alternating and direct current and undersea cables) was only about 6,000 miles, and the longest link was just 110 miles (220 kV, 220 MW), between New South Wales and South Australia.[33] This record-long tie (built to trade electricity between the two adjacent states) required a two-year permitting process, even though it goes mostly through the bush, and twenty-one months of construction.

Contrast all these accomplishments with the requirements for America's new supergrid. The country would need at least 50,000 miles of new lines, with multiple underground links from the Great Plains to the coasts each more than 1,000 or even 1,500 miles long, and capacities for each of these lines would have to be in the multiples of gigawatts, not a few hundred megawatts. The whole project would require considerable and rapid scaling up of the existing system. To think that these megaprojects could be designed, the designs approved, and the necessary rights of way obtained in a few years is to have an entirely unrealistic understanding of America's engineering capabilities, its multiple regulatory bureaucracies, and its extraordinary NIMBYism and litigiousness.

FIGURE 8-5

AMERICA'S FUTURE HIGH-VOLTAGE TRANSMISSION CHALLENGE

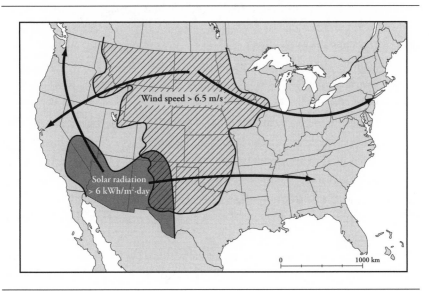

SOURCE: Author's illustration.

There is no point in fully deconstructing the Pickens plan, which would also have required a massive construction of long-distance HV lines, besides converting most of America's filling stations to dispense natural gas as well as gasoline. In October 2008, Pickens began to warn that the unfolding credit crunch would imperil the project's initial centerpiece, the 4 GW wind farm in West Texas to be built by his Mesa Power Company. In November 2008 he announced that the project would be scaled back, and by July 2009 the plan was suspended.[34] Clearly, America will not see any grand Pickensian wind-for-natural-gas swap within ten years.

And yet in comparison with the latest proposal for a rapid energy transition, both the Gore plan, and even more so the Pickens plan, are models of restraint and relative modesty. The first deals "only" with America's electricity, the other "only" with the country's electricity and cars. In contrast, Jacobson and Delucchi[35] propose to convert all of the world's energy supply to sustainable energy in just two decades by following the WWS (wind, water, and sunlight) path. Given the fact that most of the contemplated capacity in large hydrostations is already in place, their grandiose plan rests on installing 3.8 million large (each with 5 MW capacity) wind turbines and 89,000 photovoltaic and concentrated solar power plants (averaging 300 MW). They estimate the cost of all of this (excluding the requisite new transmission lines) on the order of $100 trillion.

Accomplishment of this lightning-fast extravaganza would require abandoning (except for hydro dams and HV lines) all of the world's existing energy infrastructure and erecting a brand new one by 2030. Average annual cost of this enterprise—taking into account its authors' estimate and adding the cost of extensive new transmission grids, lost capital value of the suddenly abandoned fossil-energy industries, and forgone revenue from their terminated operations—would be easily equal to the total value of the U.S. gross domestic product (GDP), or close to a quarter of the global economic product.

My verdict concerning this project's feasibility has been shared by many other life-long students of energy and could not be expressed better than by quoting just two of many scathing comments submitted to the editors of *Scientific American*, in which the Jacobson and Delucchi proposal appeared. Michael Briggs wrote: "As a physicist focused on energy research, I find this paper so absurdly poorly done that it is borderline irresponsible. There are so many mistakes, it would take hours of typing to point out all of the problems. The fact that *Scientific American* publishes something so poorly done does not speak well of the journal."[36] And Seth Dayal added: "This paper is an irresponsible piece of nonsense that would generally be found for order in the back pages of some pulp fiction magazine. The sad part is the editors for some reason chose to not only publish the claptrap but to endorse it."[37]

It is one thing when a former politician endorses an unrealistic project to boost his media presence or when an astute businessman pushes a scheme that would eventually benefit his investments—but it is an entirely different matter when one of the world's oldest science magazines lends its pages to fairy tales that any seasoned engineer and any responsible student of energy systems find grotesquely immature.

The historical verdict is unassailable. Because of the requisite technical and infrastructural imperatives and because of numerous and often entirely unforeseen socioeconomic adjustments, energy transitions in large economies and on a global scale are inherently protracted affairs. That is why, barring some extraordinary—better yet, truly heroic and entirely unprecedented—commitments and actions, none of the promises for greatly accelerated energy transitions will be realized. Moreover, during the next decade, none of the new energy sources and prime movers will make a major difference by capturing 20–25 percent of its market, either world-wide or in the United States. A world without fossil fuel combustion is highly desirable, and, to be optimistic, our collective determination, commitment, and persistence could accelerate its arrival. But getting there will be expensive and will require considerable patience. Coming energy transitions will unfold, as the past ones have done, across decades, not years.

Conclusion:
Lessons and Policy Implications

My review and deconstruction of prominent energy myths based on first principles, basic engineering realities, and simple but revealing quantifications offer lessons ranging from the obvious to the subtle. The histories of some of these myths amount to cautionary tales of misplaced zeal and exaggerated claims whose message should have been heeded a long time ago; other myths, which describe processes that could be useful or desirable if developed on appropriate scales and introduced according to realistic timetables, show the folly of uncritical advocacy, hasty implementation, and enthusiasm unchecked by a closer examination of complex realities.

To summarize the lessons from my review of the individual energy myths, I will discuss their particular implications and at the same time point out the broader, generic applicability of such lessons. These generic lessons are particularly important, given the number of new proposals, suggestions, and arguments now being offered—some of them by the new administration—to ease and perhaps even solve America's considerable energy challenges.

Electric Vehicles

By far the most important specific lesson to be learned from plans for a large-scale market penetration by electric vehicles is that such a seemingly revolutionary prime-mover shift would not amount to any significant primary energy savings. If all the additional electricity needed by these vehicles were generated either by using the existing infrastructure more intensively—that is, using coal- and natural gas–fired plants with higher capacity factors—or by adding new capacities operating with similar efficiencies, there would be a major reduction in crude oil demand but no notable weakening of the U.S. dependence on fossil fuels generally.

Even if all automakers were to meet all their current plans for the introduction of all-electric automobiles,[1] these cars would still account for less than 2 percent of the world's passenger vehicles on the road by 2015. Today, although there are better technical reasons and stronger economic and environmental incentives to bring about a mass adoption of electrics than a century ago, it would be naïve to expect that the makers and users of internal combustion engines will simply abandon their machines.

Major reductions in fossil fuel combustion and carbon dioxide emissions would be realized only if the additional electricity needed to power the new vehicles were to come from the most efficient combined-cycle generation or from renewable sources. That would, of course, require large infrastructural investment in wind turbines and photovoltaics that could not be put into place in a matter of a few years. We could more quickly reduce current dependence on refined fuels by relying on highly efficient gasoline vehicles, by retiring as quickly as possible the least-efficient vehicles,[2] and by continuing to diffuse already proven and reliable hybrid drive designs.

The saga of electric cars, more than a century old, also offers a number of important generic lessons for policymakers. First, during the earliest stages of a technical innovation, it is not only difficult but often simply impossible to discern which option will eventually dominate. In 1900, expert consensus was behind the electrics, but eight years later Ford's Model T turned them into a lingering curiosity. Likewise, in 2000 the experts were betting on fuel cell cars, but within a few years the focus was on hybrids, until it shifted, once again, to the electrics. Similar examples abound in other energy sectors; a prominent case is the failure of breeder reactors noted earlier.

Second, the eventual fate of a particular option often depends on the availability of suitable infrastructure. Thus, even if better batteries had been available before World War I, it still would have been easier to build a network of gasoline stations than to provide extensive and reliable recharging during the pioneering era of inefficient electricity generation and limited transmission. Analogically, it would be easier today to burn natural gas more efficiently in combined-cycle arrangements added at existing thermal plant locations and use the additional electricity in hybrid or electric car drives than it would be to follow the Pickens plan to replace natural gas–fueled generation by wind turbines and use the saved natural gas to power cars.

Third, an eventually dominant option does not have to be superior on all counts. Compared to the electrics, pre–World War I gasoline-fueled cars were difficult to start, noisy, and polluting—but their mass production made them *affordable* and *reliable*. And those two qualities will decide the eventual fate of many future innovations.

Fourth, the reintroduction of the electrics faces a challenge common to all engineered systems: the considerable inertia of the dominant prime mover. As a winning option diffuses massively,[3] as it acquires a vast infrastructural support,[4] and as both the producers and the users become knowledgeable about and comfortable with using and improving the dominant technique, the incentive to remain with a known and proven—even if not optimal—choice is strong.

Finally, mass-scale adoption of a new technique may not be best served by perpetuating an aura of injured superiority, misunderstood excellence, and unappreciated technical elegance, or by making claims of inevitable though long-deferred dominance. Likely failures or missed milestones of market penetration may then lead to a reverse overcompensation—that is, to a hasty dismissal of any future realistic possibilities of these new techniques. Electrics have been burdened by all of these—as have fuel cells, decentralized electricity generation, and the hydrogen economy. Meanwhile, our dependence on fossil fuels continues to be nearly as strong as ever.

Nuclear Power

The unrealistic—and ultimately unmet—expectations expressed for nuclear power offer a most important generic lesson for anyone considering the claims, often no less exaggerated, made today on behalf of new energy conversions yet to be commercialized on a large scale, like photovoltaics or fuel cells, or on behalf of already proven techniques in the early stages of significant commercial diffusion, like wind turbines or hybrid cars. But a no less important lesson with a broad applicability can be learned from America's experience with nuclear fission, which constitutes a nearly perfect example of what I have called a successful failure of technical innovation.

In such a case, a new technique conquers a substantial share of its market and proves reliable and economical, but both because its importance falls far short of initial and unrealistic expectations and because it has not resolved some of its long-term operational challenges, it is seen as a questionable undertaking whose further expansion is perhaps best avoided. A repetition of this experience could be the fate of one or more of the new forms of energy conversion that are now extolled as perfect long-term supply solutions. Will the current uncritical promotion of wind-powered electricity generation survive the inevitable challenges presented by the massive extension of high-voltage transmission and by the integration of generators with different load capacities? How will the future of PV generation be affected by the failure of today's unrealistic predictions of dramatically declining unit cell prices?

The failures of time-specific, long-range capacity forecasts for nuclear generation should remind us that all detailed quantitative predictions of technical advances on a multidecadal scale are highly questionable, if not inherently futile. The best we can do is to discern an important trend and perhaps its most likely intensity and stay away from detailed forecasts. At best, they will widely miss their targets; at worst, they will be quite risible. Unfortunately, this lesson is commonly ignored, as not only promoters with economic or ideological stakes in specific techniques but also many governments continue to call for unrealistic performance targets or publish patently exaggerated forecasts of future market shares to be captured by new energy conversions.

The principal specific lesson from the nuclear past is clear: There has never been any chance of having nuclear electricity too cheap to meter. That claim was never anything more than a hyperbolic, inspirational phrase unsupported by any facts. But policy conclusions based on the history of nuclear generation are no less obvious, and they are based on qualified success rather than on failure. Well-managed nuclear fission has proved to be a significant and reliable contributor to the modern electricity supply; moreover, even at a relatively high price, it may yet turn out to be one of the best choices to moderate the rate of anthropogenic climate change, because we know how to design and build nuclear power plants that can be operated safely and with very high load factors. Electricity produced by nuclear fission is not carbon free. Fossil fuels are used to produce materials

FIGURE C-1

CO$_2$ EMISSIONS FOR DIFFERENT FORMS OF PRIMARY ENERGY

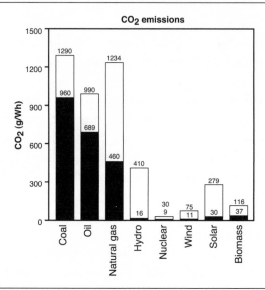

SOURCE: Plotted from data in International Atomic Energy Agency (IAEA, 2001).
NOTE: Black and white bars represent, respectively, minima and maxima for different energy conversion efficiencies.

needed for building nuclear stations, and a large share of electricity used during the enrichment of uranium comes from fossil-fueled generation. Its carbon intensity, however, is minuscule compared to that of other forms of primary energy (see figure C-1).

Consequently, no rational long-range energy plan of any major modern economy should exclude the nuclear option; the debate should be about the best way to proceed, not about whether to proceed at all. At the same time, all such debates must take into account the challenge of those notoriously irrational risk perceptions that affect not only the launching of new nuclear construction programs but also the siting and maintenance of permanent disposal sites for radioactive wastes. Unfortunately, there has been no serious, concerted effort—akin to the decades-long campaigns that decisively convinced much of the public of the perils of smoking—aimed at placing the risks of nuclear electricity generation in their proper context. This is a failure of both science and policy.

Soft-Energy Conversions

Two obvious generic lessons arise from the failure of soft-energy conversions to claim significant shares of energy services in modern societies during the past few decades. The first is that the pursuit of any long-term energy goals should not be animated by ideology; the second is that such a pursuit should not elevate a single class of techniques or managerial approaches to an all-encompassing means of creating the preferred energy supply system. Fundamental transformations of energy supply and use will have their inevitable socioeconomic and political consequences, but energy supply and use should not be primarily a tool of some desired social transformation. And while no sensible choices should be excluded from the pursuit of long-term transformation of the energy supply, no single choice should enjoy an *a priori* preference.

Other generic lessons are similar, or virtually identical, to those derived from the experience with nuclear generation. First, true believers—those biased advocates of particular approaches—cannot be expected to offer objective appraisals of the solutions they advocate. Consequently, their proposals should not be accepted just because of assurances that they possess many advantages over other approaches, that they will promote some desirable social or environmental end, or that they can be tools of socioeconomic, political, or environmental transformation.

Second, claims about the future performance of techniques and processes in their early stages of commercialization, and about the rates of their likely practical diffusion, must be critically examined and questioned. At the same time, exaggerated claims and failed forecasts do not justify a blanket dismissal of those underperforming or failing approaches. Their more rational introduction and less expansive diffusion may still turn out to be a highly valuable component of an overall energy solution.

The soft-energy experience vividly demonstrated that the diligent pursuit of rational energy solutions must be firmly grounded in an understanding of complex realities—involving available resources and the engineering imperatives of conversion techniques and infrastructural requirements. It must also take into account prevailing distribution and consumption patterns, as well as the accessibility and reliability of the desired energy services—heat, motion, or light. Unbiased examination of these realities and requirements shows a number of significant mismatches between the supply that can be delivered by decentralized, small-scale conversions—with their overwhelmingly low power densities, inherently limited unit capacities, and often suboptimal conversion efficiencies—and the demand of modern, urban, and industrial areas in general and megacities in particular.

Peak Oil

An obsession with the actual timing of peak oil perfectly illustrates the futility of assigning a date to predicted future events. Such dates are bound to be wrong, and those who take them seriously are bound to be misled. This obsession also illustrates the common mistake of focusing on what is, in reality, a matter of secondary importance. In the case of peak oil, we should anticipate the inevitable but impossible-to-date decline of inexpensive liquid oil resources by shifting gradually to other energy sources, so that the peak output, whenever it comes, is no more noteworthy than were the past peak outputs of fuel wood or coal production.

Another important generic lesson comes from the way the aftermath of an imminent oil peak has been portrayed: not just as an event with dire economic consequences, but as the demise of modern civilization. This book has aimed at criticizing assorted myths and misconceptions, and in doing so has mostly had to correct excessively positive or unjustifiably enthusiastic expectations and interpretations. This case makes clear, however, that an opposite kind of correction is sometimes necessary. The world will not end when global crude oil extraction stabilizes or when it eventually begins its long decline, as it did not end when we stopped the aggressive pursuit of nuclear power or realized that breeder reactors would not be our salvation. Nor will it end if we do not convert to a hydrogen economy, if we do not pursue the biofuel option, if we do not try to make every car run on electricity, or if we do not sequester compressed carbon dioxide underground. No individuals and no collective deliberations are clever enough to offer a blueprint to be followed for decades. All we can do is to work with aspirations, which always require numerous corrections and adjustments, and to anticipate reversals. And, undoubtedly, peak-oilers could have a helpful and positive influence on energy debates if they were less catastrophic in their visions and used some of their correct arguments to focus our efforts on practical solutions rather than on disseminating the end-of-the-world-as-we-know-it scenarios.

Carbon Sequestration

By far the most important generic lesson taught by a critical examination of plans for large-scale carbon sequestration is that avoiding or minimizing an undesirable environmental impact is far superior to any efforts aimed at neutralizing that impact once it has already taken place. We should embrace the superiority of the first approach as a key engineering and management maxim. The proponents of mass-scale carbon sequestration, however, dismiss it by maintaining that further large increases in emissions are inevitable, and hence that the only practical options for reducing their impacts is their capture and storage. This conclusion appears to follow a realistic assessment with a rational response, but in fact neither claim can withstand critical examination.

A closer examination of possible outcomes makes clear that the inevitability claim must be qualified. Affluent nations, the largest per-capita emitters, could moderate and manage their energy use in a way that would intensify the ongoing decarbonization of their economies and result in absolute declines of overall emissions rates (see figure C-2). After all, the past two decades have already brought no, or only very slight, increases in per-capita energy use in many rich countries, and combinations of aggressively promoted rationalization, driven by innovation and constructive taxes, could turn this into a slowly declining trend. Rapidly modernizing low-income countries, moreover, doubtlessly could achieve substantial economic gains without replicating many undesirable choices—such as in the design of buildings and transportation systems—that have been made by affluent nations. We should thus try, earnestly and with some perseverance, to exhaust all avoidance strategies before we commit to any massive (and potentially risky) underground storage of many billions of tons of CO_2.

The claim regarding the practicability of massive CO_2 sequestration belongs to that large category of promises that rely on insufficient information and experience. Mastery of the constituent processes needed to operate

FIGURE C-2

DECLINING CARBON INTENSITY OF THE U.S. ECONOMY, 1950–2010

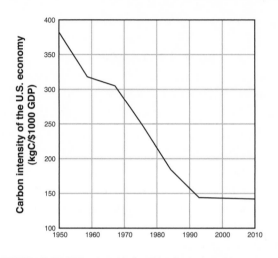

SOURCES: Calculated from data in Marland et al. (2007) and U.S. Bureau of Economic Analysis (2008).

sequestration systems (such as expertise in compression, pumping, and pipeline transportations of liquids and gases), along with the theoretical assessments of feasibility based on preliminary appraisals of suitable sedimentary formations and several years of monitoring small-scale pilot projects, do not add up to an understanding sufficient to justify grandiose plans for an economically acceptable and technically reliable enterprise operating with annual throughputs rivaling those of the world's largest material-handling industries. If serious steps toward carbon sequestration are taken soon, they will represent yet another failure to appraise a promising technique carefully before committing to its large-scale commercial adoption.

Crop-Based Ethanol

Rapid expansion of crop-based ethanol production is an unfortunate but all-too-perfect example of a new energy choice driven by overemphasis on a few positive aspects of a technical innovation, combined with inexplicable neglect of many of its negative consequences. A steady income for corn growers, investment in domestic energy production and technical innovation, and a reduction in oil imports cannot make up for the enormous environmental impacts of expanded and intensified fuel-crop cultivation, for higher national and global food prices, or for the enormous subsidies that will be required—especially as crop-derived ethanol can have only a marginal effect on the quest for a higher degree of energy self-sufficiency.

Indeed, the costs of crop-derived ethanol greatly outweigh the benefits, as the economic, social, and environmental costs will greatly surpass all those relatively minor and inevitably ephemeral benefits in the long run. Perhaps the most important generic lesson to be derived from the recent U.S. experience with ethanol is that we need to separate completely all decisions on long-term energy policies from any short-term corporate interests[5] and from all dubious partisan promises, especially those made as a part of political campaigns, where unrealistic expectations of rapid renewable ascendance ignore many environmental, engineering, and economic realities.

Finally, the embrace of biofuels as strategic game changers or as tools of green politics illustrates the dangers of concentrating on dubious secondary solutions while ignoring many factors that are incomparably more important and more decisive. America's long-term strategic posture would be far better served by sound fiscal and consumption policies—that is, balanced federal and state budgets and an end to runaway trade deficits—and by a strong commitment to continuous technical innovation that could realistically double the average efficiency of today's vehicle fleet than by spending billions to convert Midwestern corn to ethanol.

Wind-Powered Electricity Generation

As in ethanol's case, recent exaggerated expectations for rapid, reliable, and sustained contributions made by wind-powered generation are based on a selective reading of the evidence. The resource is undoubtedly large, but the power that could be harnessed economically (that is, an equivalent of wind reserves) is considerably smaller, most likely less than 10 percent of the theoretical capacity. Moreover, windy sites best suited for the most profitable generation have a highly uneven spatial distribution, a reality that restricts immediate large-scale development of wind power to regions that already have fairly good long-distance high-voltage interconnections with areas well supplied by other sources of electricity.

In all other instances, large—that is, GW-scale—wind projects will be able to proceed only once the requisite HV links, or grids, are in place, but as these new links need substantial upfront capital investment, and as their planning and approval process takes many years (even without obstructive litigation), their progress is unlikely to match the goals for the shares of wind-powered electricity that have been so readily posted by many governments for 2020 or 2030. Nor should we underestimate the consequences of a mismatch between the electricity demands of modern societies and typical load factors of wind-powered generation.

Round-the-clock electronic controls and communications have become increasingly demanding components of modern electricity use, and long-range plans in countries ranging from the United States to China envisage higher reliance on electric trains and cars. These uses will increase the base load of electricity generating systems whose demand is today provided most reliably by nuclear power reactors with load factors commonly in excess of 90 percent—and not by wind-powered generation with its low loads of 20–25 percent. Properly sited, well-engineered wind farms built as parts of well-interconnected grids should have an important place in the coming energy transition. Such a mission is challenging enough without raising unrealistic hopes that this new conversion cannot fulfill.

Energy Transitions

Ignoring the lessons of energy transitions has been a common transgression committed by overenthusiastic promoters of new fuels, such as crude oil from Canada's oil sands and liquid biofuels, and of new energy conversions, such as automotive fuel cells, central solar power, or commercial electricity generation by assorted wave-harnessing devices. Without exception, these enthusiasts have come up against sobering realities. Expensive, and environmentally taxing, extraction of oil from Alberta's oil sands produced about 40 percent of Canada's crude oil in 2008, but in global terms this amounts to less than 2 percent. In addition, uncertainties regarding future production costs make any long-range extraction forecasts nothing but guesses. An initially uncritical embrace of crop-based biofuels and some grandiose plans in 2005 and 2006 were followed by a wave of critical analyses exposing the excessive cost of these fuels and their deleterious impacts on the environment and on food prices.

A particularly telling case in point is the fate of cars powered by fuel cells, and nothing illustrates that better than the history of Ballard Power Systems of Burnaby, British Columbia. The company was established in 1979 by Geoffrey Ballard, and in its quest for hydrogen-powered transportation it became the paragon of automotive fuel cell promise. The founder took it public in 1993 on the Toronto Stock Exchange and in 1995 on the NASDAQ. Its stock began to rise in 1997; then, as the unrealistic expectations about the imminent arrival of fuel cell cars mounted, it surpassed C$200/share in early 2000, and assorted dignitaries trekked to Vancouver to drink the pure water exhaust dripping from the tailpipe of a demonstration hydrogen bus using Ballard's fuel cell. But shortly afterward, the realities intervened, and before the end of 2000 the stock began a long slide that brought it to about C$3/share by the end of 2008 (see figure C-3), and it is still there in 2010. The company has totally withdrawn from any development of hydrogen-fueled propulsion to concentrate on fuel cells for forklifts and backup electricity generation.

The history of energy transitions strongly suggests that no grandiose plans aimed at large-scale and rapid changes in the composition of primary energy supply or at accelerated commercial adoption and widespread diffusion of new conversion techniques have been particularly successful. Analyses of energy transitions in the world's major economies have one notable thing in common: remarkable persistence of primary energy supply patterns during the past two generations. Wishful thinking, pioneering enthusiasm, and belief in the efficacy of seemingly superior solutions are not enough to change the fundamental nature of gradually unfolding energy transitions, be they the shifts to new fuels, to new modes of electricity generation, or to new prime movers. Adoption of these innovations is nearly always tied to major infrastructural developments that require large capital investment; moreover, it inevitably confronts environmental, legal, and organizational complications and can be hindered by irrational perceptions of risk.

FIGURE C-3

VALUE OF BALLARD POWER SYSTEMS SHARES ON
TORONTO STOCK EXCHANGE, 1994–2008

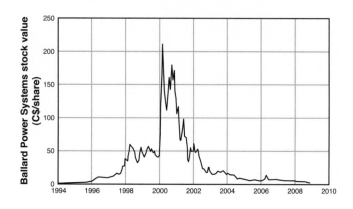

SOURCE: Based on a chart that can be created at the Toronto Stock Exchange website (http://www.
tmx.com, symbol BLD-T).

A Quick Summation

I close with a handful of terse summations that capture the reasons for the
prevalence and persistence of energy myths and that offer some advice
about how to resist their allure.

First, distrust any strong, unqualified claims regarding the pace, timing,
and extent of future adoption of new energy sources or the diffusion and
performance of new energy conversion techniques. Perhaps the most obvious corollary of this proposition is to avoid any temptation to fashion and
promote grand designs that promise profound changes to be accomplished
by a certain date. New arrangements of energy sources and conversions
arise in ways that are impossible to forecast in detail, as they include rapid
advances but also some remarkable failures and retreats.

Second, do not underestimate the persistence and adaptability of old
resources (remember that coal is still more important globally than natural
gas) and established prime movers, particularly those that have been around
for more than a century, including steam turbines and internal combustion

engines. Recall that the latest incarnations of the internal combustion engine, the new DiesOtto machines, have the potential to be more efficient than the best hybrid drives on today's market.

Third, do not uncritically embrace unproven new energies and processes just because they fit some preconceived ideological or society-shaping molds. Wind turbines or thin-film solar cells may seem to be near-miraculous forms of green salvation, ready to repower America within a decade. But ours is a civilization that was created by fossil fuels, and its social contours and technical foundations cannot be reshaped in a decade or two.

Fourth, be mindful of extensive, and often very expensive, infrastructural requirements that must be put into place before any new ways of energy supply and use can be widely adopted. Remember that even the rise of that most iconic of all rapid technical innovations, microprocessor-based electronics, could not have taken place without the previous establishment of an enormous infrastructure for electricity generation (overwhelmingly fossil-fueled) and transmission.

Fifth, remember that energy transitions are inherently prolonged affairs lasting decades, not years. Consequently, avoid, by all means, the tendency to confuse energy innovations with advances in computing power. Energy innovations are not subject to Moore's law, and we cannot keep doubling the efficiency of our energy use or halving the cost of those conversions in a matter of eighteen to twenty-four months.

Finally, and, perhaps unfortunately: Let us not have any illusions that myths can be uprooted simply by appealing to reason. The Roman playwright Terence knew that more than twenty-one centuries ago, and that is why I chose his words as this book's epigraph as well as its ending: "*Homines libenter quod volunt credunt*" (Men believe what they want to). It may be true, but it is hardly the best foundation for rational energy or any other policies.

Notes

Introduction

1. The rapidly increasing capture of wind by large turbines and of solar radiation by photovoltaic cells still makes only a negligible contribution globally.

2. By 1986 their annual average fell to less than $15/barrel, or about $30/barrel in 2008 dollars (BP 2009).

3. Energy independence is an unrealistic proposition, given that all major Western economies trade different forms of energy.

4. World Trade Organization (WTO) 2009.

5. By 2007, those designs accounted for nearly 45 percent of the total U.S. vehicle count (U.S. Department of Transportation, Bureau of Transportation Statistics 2007).

6. All Western countries now have lower oil intensity—that is, they need considerably less oil to produce a dollar of gross domestic product—than they did in the early 1980s. In 2008 the U.S. economy needed (per constant GDP dollar) only 48 percent as much oil as it did in 1980.

7. The OPEC basket price was introduced in June 2005 and is made up of prices for twelve crude oils from the organization's member countries.

8. Kurzweil and Meyer 2003.

9. Gordon E. Moore was the cofounder of Intel. "Moore's Law" refers to a concept he introduced in a 1965 paper, describing a longstanding trend whereby the processing power of microchips doubles every two years. See Moore (1965) and Smil (2008b).

10. Roush 2006.

11. Seife 2008.

12. Power-Technology.com 2009.

13. Criswell 2000.

14. OTEC News 2009.

15. Olah et al. 2006.

16. Parkins 2006.

17. Rudin 2004; Herring 2006; Polimeni et al. 2007.

18. Jevons 1865, 140; emphasis in the original.

Part I: Lessons from the Past

1. Shnayerson 2007, 9.

Chapter 1: The Future Belongs to Electric Cars

1. Smil 2004.
2. Ibid.
3. Ibid.
4. Burwell 1990.
5. Kirsch 2000.
6. McShane 1997.
7. Ford 1922, 34.
8. Josephson 1959.
9. Smil 2004.
10. Imbrecht 1995.
11. Lazaroff 2001.
12. Paine 2006.
13. Kirsch 2000.
14. *Electrifying Times* 2007.
15. EV Innovations Inc. (EVII) 2009.
16. Commuter Cars Corporation n.d.
17. Shnayerson 2007, 1.
18. Tesla Motors n.d.
19. Ibid.
20. Berdichevsky et al. 2006.
21. Voelcker 2007.
22. Bennett 2008.
23. Jaffe 2007, 44.
24. General Motors 2009.
25. Farah 2008.
26. Renault 2009.
27. Better Place 2009.
28. Tollefson 2009.
29. IHS Global Insight 2009.
30. European Federation for Transport and Environment (EFTE) 2009.
31. Deutsche Welle 2009.
32. Eberhard and Tarpenning 2006.
33. EFTE 2009.
34. U.S. Department of Energy, Energy Information Administration 2009.
35. Kintner-Meyer et al. 2007.
36. U.S. Department of Energy, Energy Information Administration 2009.

37. EFTE 2009.
38. Ibid., 5.
39. International Energy Agency (IEA) 2008.
40. Daimler 2009.
41. Nazri and Pistoia 2004.

Chapter 2: Nuclear Electricity Will Be Too Cheap to Meter

1. Strauss 1954, 5.
2. Lilienthal 1959, 21.
3. Lowen 1987.
4. Rockwell 1992.
5. Atkins 2000.
6. Cowan 1990.
7. Seaborg 1971.
8. Meier 1956; Oak Ridge National Laboratory (ORNL) 1968; Seaborg 1968.
9. Seaborg 1972, 34.
10. Seaborg and Corliss 1971.
11. Rose 1974, 351.
12. Bethe 1977, 59.
13. Rossin and Rieck 1978, 582.
14. Olds 1982.
15. Feldman et al. 1988.
16. Semenov and Oi 1993.
17. International Atomic Energy Agency (IAEA) 2008.
18. U.S. Atomic Energy Commission, Division of Technical Information 1971.
19. Weinberg 1973, 18.
20. Creagan 1973, 16.
21. Murphy 1974.
22. Von Hippel and Jones 1997.
23. Japan Nuclear Cycle Development Institute (JNCDI) 2000.
24. International Atomic Energy Agency (IAEA) 2008.
25. Ibid.
26. Ibid.
27. It is not entirely carbon-free, because fossil fuels are still needed to mine and process uranium and to produce the materials needed to build power stations.
28. Moore 2006.
29. Lovelock 2006.
30. Van der Zwaan 2002.
31. Marshall 2005.
32. Marris 2006.
33. Charles 2007.

34. Hohenemser et al. 1977.

35. Denning 1985; Walker 2004.

36. Hohenemser 1988; Nuclear Energy Agency (NEA) 2002.

37. Nuclear Energy Agency (NEA) 2002.

38. Garwin and Charpak 2001; Chapin et al. 2003.

39. Weinberg 1994, 17.

40. Organisation for Economic Cooperation and Development (OECD) 2000; International Atomic Energy Agency (IAEA) 2006.

41. Teller et al. 1996.

42. *China Daily* 2009.

43. U.S. Department of Energy, Energy Information Administration 2009.

44. Ferguson et al. 2009; Brumfiel 2008.

45. Smil 2006, 63.

46. Bupp and Derian 1978; Weinberg 1994; Cohn 1997; Makhijani and Saleska 1999.

47. Nuclear Information and Resource Service (NIRS) 1999.

Chapter 3: Soft-Energy Illusions

1. Lovins 1976, 65.

2. Ibid., 65–66.

3. Ibid., 77–78.

4. Ibid., 82.

5. Ibid., 82.

6. Schumacher 1973.

7. Lovins 1992, 9.

8. As explained earlier, the 1990s saw their demise.

9. U.S. Department of Energy, Energy Information Administration 2009.

10. Hawken 1997.

11. Chevalier 2000.

12. Rocky Mountain Institute (RMI) 2007.

13. InterTechnology Corporation 1977.

14. Stobaugh and Yergin 1979.

15. Hayes et al. 1979.

16. U.S. Department of Energy, Energy Information Administration 2009.

17. Johansson and Steen 1978.

18. I use the adjective *Maoist* to describe the period when the country's policies were dominated by the whims of the Great Helmsman. Although he died in September 1976, true de-Maoization began only once Deng Xiaoping's economic reforms—tentatively introduced in 1979—gathered momentum in the mid-1980s.

19. Smil 1999.

20. Smil 1988.

21. Huang and Chang 1980.
22. Smil 2004.
23. Carin 1969.
24. Lovins 1976, 77.
25. Japan's Kansai region, the U.S. Northeast between Boston and Washington, and China's Zhujiang Delta are examples.
26. Lovins 1978, 511.
27. Lovins 1976, 9.

Part II: Myths in the Headlines

1. British Petroleum (BP) 2009.

Chapter 4: Running Out: Peak Oil and Its Meaning

1. Ivanhoe 1995, 5.
2. Duncan 2000.
3. Ibid.
4. British Petroleum (BP) 2009.
5. Hubbert 1956.
6. Hubbert 1969, 192.
7. British Petroleum (BP) 2009.
8. Ibid.
9. Brandt 2007, 3084.
10. Workshop on Alternative Energy Strategies (WAES) 1977.
11. Flower 1978, 42.
12. Central Intelligence Agency (CIA) 1979.
13. Deffeyes 2004.
14. Nehring 2006a; 2006b.
15. Nehring 2006c, 51.
16. Adelman 2004, 18.
17. U.S. Geological Survey 2000.
18. Cambridge Energy Research Associates (CERA) 2006.
19. Government of Alberta 2010.
20. Canadian Association of Petroleum Producers (CAPP) 2009.
21. British Petroleum (BP) 2009.
22. Ibid.
23. Odell 2006.
24. Kerr 2007.
25. El-Badri 2008.
26. Tillerson 2008; Al-Falih 2009.
27. Simmons 2009.

28. U.S. Census Bureau 2009.
29. White House 2009.
30. British Petroleum (BP) 2009.
31. Smil 2008.
32. Jochem 2002.

Chapter 5: Sequestration of Carbon Dioxide

1. Marland et al. 2007.
2. Wigley and Schimel 2000.
3. Potter et al. 2003.
4. Fan et al. 1998.
5. Potter et al. 2003.
6. Nabuurs et al. 2003.
7. Beer et al. 2006.
8. Stephens et al. 2007.
9. Nemani et al. 2002.
10. Percy et al. 2009
11. De Vries et al. 2006.
12. Janssens et al. 2005.
13. United Nations Food and Agriculture Organisation 2003.
14. Schlesinger 2000; Lal et al. 2004.
15. Loya et al. 2003.
16. Bellamy et al. 2005.
17. Heath et al. 2005.
18. Trumbore and Czimczik 2008.
19. Walter et al. 2006.
20. Lehmann et al. 2006; Lehmann 2007.
21. Lehmann et al. 2006.
22. Martin and Fitzwater 1988.
23. Coale et al. 2004.
24. Tréguer and Pondaven 2000.
25. Informationsdienst Wissenschaft (IDW) 2009.
26. Strong et al. 2009.
27. Smith et al. 2005.
28. Carey et al. 2001.
29. Marchetti 1989.
30. Jayaraman 2007.
31. Goldberg et al. 2008.
32. Kelemen and Matter 2008.
33. Lackner et al. 1999; Keith 2009.
34. Rochelle 2009.

35. Dooley et al. 2009.

36. U.S. Department of Energy 2008.

37. Rai et al. 2008.

38. Chandler 2007.

39. Intergovernmental Panel on Climate Change (IPCC) 2006.

40. Ibid.

41. International Energy Agency (IEA) 2006.

42. Intergovernmental Panel on Climate Change (IPCC) 2006.

43. Kharaka et al. 2006.

44. Gilfillan et al. 2009

45. Citizens against CO_2 Sequestration 2009.

46. Leakage from a CO_2 pipeline could lead to an accumulation of denser-than-air gas in low-lying areas and to accidental suffocation.

47. Wilson et al. 2008.

48. National oil companies, some within OPEC, control the rest.

49. Greenpeace International 2008.

50. Intergovernmental Panel on Climate Change (IPCC) 2006.

51. Chu 2009, 1599.

Chapter 6: Liquid Fuels from Plants

1. Solomon et al. 2007.

2. Smil 2005.

3. Lugar and Woolsey 1999.

4. Fairless 2007; Sanderson 2009.

5. British Petroleum (BP) 2009.

6. U.S. Department of Energy, Energy Information Administration 2009; British Petroleum (BP) 2009.

7. Solomon et al. 2007.

8. Koplow 2006; Steenblik 2007.

9. Pimentel 2003.

10. U.S. Department of Agriculture 2004.

11. Kim and Dale 2002.

12. Durante and Miltenberger 2004; Pimentel and Patzek 2005; U.S. Department of Agriculture 2004.

13. Howarth and Bringezu 2009.

14. U.S. Department of Agriculture 2007.

15. Cassman et al. 2002.

16. Rabalais 2002.

17. Guru and Horn 2000.

18. Campbell et al. 2009.

19. Howarth and Bringezu 2009.

20. Brazil, Government of the State of São Paulo, Secretariat of the Environment 2004.
21. De Oliveira et al. 2005.
22. Patzek and Pimentel 2005.
23. Brazil, Ministério de Agricultura, Pecuária e Abastecimento 2007.
24. Rípoli et al. 2000.
25. Sparovek and Schnug 2001.
26. Runge and Senauer 2007.
27. U.S. Department of Agriculture 2007.
28. U.S. Congressional Budget Office 2009.
29. Ferrett 2007.
30. Von Lampe 2006.
31. U.S. Department of Energy 2007.
32. Khosla 2006.
33. Ibid., 72.
34. Smil 1999.
35. Graham et al. 2007.
36. Pordesimo et al. 2004.
37. Blanco-Canqui et al. 2007.
38. Kadam and McMillan 2003.
39. Thomas 2003.
40. Richey et al. 1980; Shinners et al. 2007.
41. Shinners et al. 2007.
42. Perlack et al. 2005, 19.
43. Service 2007; Stephanopoulos 2007; Himmel et al. 2007.
44. Solomon et al. 2007.
45. Service 2007.
46. Kwok 2009.
47. Stephanopoulos 2007.
48. Parrish and Fike 2005.
49. Raghu et al. 2006.
50. Felix and Tilley 2009.
51. Kondamudi et al. 2008.
52. Reyes and Sepúlveda 2006.
53. Tacon 2004.
54. International Energy Agency (IEA) 2005, 39.
55. Schouten 2005.
56. Ibid.
57. Gillis 2006.
58. Sustainable Mobility Project (SMP) 2004.
59. Mascarelli 2009.
60. Giampietro and Mayumi 2009.

Chapter 7: Electricity from Wind

1. Archer and Jacobson 2005.
2. American Geophysical Union (AGU) 2005.
3. Archer and Jacobson 2003.
4. Brown 2003, 1.
5. Lu et al. 2009.
6. Smil 1994.
7. ENERCON 2009.
8. Ibid.
9. Archer and Jacobson 2005.
10. Peixoto and Oort 1992; Lorenz 1976.
11. Sky WindPower 2009.
12. Magenn Power 2009.
13. Vance 2009.
14. Gustavson 1979, 14.
15. Archer and Jacobson 2005.
16. Pasqualetti et al. 2002.
17. Marris and Fairless 2007.
18. U.S. Department of Agriculture, Forest Service 2005.
19. Jacobson and Masters 2001a.
20. DeCarolis and Keith 2001.
21. Awerbuch 2006.
22. Archer and Jacobson 2005.
23. International Commission on Large Dams (ICOLD) 1998.
24. McGowan and Connors 2000.
25. Keith et al. 2004; Keith and Adams 2009.
26. Boccard 2009.
27. Smil 2008a.
28. Elvidge 2004.
29. British Wind Energy Association (BWEA) 2005, 4.
30. Deutsche Energie-Agentur (DENA) 2005.
31. European Wind Energy Association (EWEA) 2007.
32. Østergaard 2008, 1459.
33. Nord Pool 2009.
34. Archer and Jacobson 2005.
35. National Renewable Energy Laboratory (NREL) 2003.
36. Jacobson and Masters 2001b.
37. Schewe 2007.
38. Cavallo 1995.
39. European Wind Energy Association (EWEA) and Greenpeace 2004.
40. Domrös and Peng 1988.

41. Global Wind Energy Council (GWEC) 2009; U.S. Department of Energy, Energy Information Administration 2009.

42. Pacala and Socolow 2004.

43. European Wind Energy Association (EWEA) and Greenpeace 2004.

44. Greenpeace International and EREC 2007.

Chapter 8: The Pace of Energy Transitions

1. Gibson et al. 2008.

2. Smith et al. 2003.

3. Pickens 2008.

4. Gore 2008.

5. Ibid.

6. U.S. Department of Energy, Energy Information Administration 2009.

7. British Petroleum (BP) 2009.

8. Nixon 1974.

9. Carter 1979.

10. Smil 2010.

11. British Petroleum (BP) 2009.

12. Ibid.

13. Ibid.

14. Smil 2010.

15. Bunker 1972.

16. Smil 2010.

17. Smil 2006.

18. Hilyard 2008.

19. British Petroleum (BP) 2009; Canadian Association of Petroleum Producers (CAPP) 2009.

20. International Energy Agency (IEA) 2009.

21. U.S. Department of Energy, Energy Information Administration 2009.

22. Ibid.

23. Ibid.

24. American Society of Civil Engineers (ASCE) 2009.

25. Gore 2008.

26. See the introduction.

27. Intel 2007, 2010.

28. Solarbuzz 2009a.

29. National Renewable Energy Laboratory (NREL) 2009.

30. Solarbuzz 2009b.

31. Gore 2008.

32. Hampton et al. 2007.

33. ABB 2008.

34. *Power* 2009.
35. Jacobson and Delucchi 2009.
36. Briggs 2009.
37. Dayal 2009.

Conclusion: Lessons and Policy Implications

1. This is a most unlikely assumption, given the parlous state of America's carmakers, the difficulties faced by the European car industry, and the considerable challenges confronting Japanese and Korean car companies.

2. These least efficient vehicles account for a disproportionate share of gasoline consumption.

3. There are now nearly 300 million gasoline-fueled vehicles in the United States and some 800 million worldwide.

4. The United States alone has nearly 120,000 gasoline filling stations, and nearly half the output of its refineries is automotive fuel.

5. In the case of ethanol, these involve massive subsidies such as those received by Archer Daniels Midland and others from the federal government.

References

ABB. 2008. Murraylink—The World's Longest Underground Power Link. http://www.abb.com/cawp/gad02181/c1256d71001e0037c1256a4e00266978.aspx (accessed January 5, 2010).

Adelman, M. A. 2004. The Real Oil Problem. *Regulation* 27 (Spring): 16–20.

Al-Falih, K. A. 2009. Powering Prosperity, Enabling Growth: Saudi Aramco's Perspective on Global Energy Security. May 6. http://www.saudiaramco.com/irj/portal/anonymous?favlnk=%2FSaudiAramcoPublic%2Fdocs%2FNews+Room%2FSpeeches&ln=en#reqEvent=Speeches&reqLang=EN&selYear=2009&selSpeaker=&launchUrl=%2FSaudiAramcoPublic%2FSpeeches%2F2009_Khalid+A.+Al-Falih_EN_May06.html (accessed January 5, 2010).

American Geophysical Union (AGU). 2005. Global Wind Map May Provide Better Locations for Wind Farms. AGU press release no. 05-14. May 16. http://www.agu.org/news/press/pr_archives/2005/prrl0514.html (accessed January 5, 2010).

American Society of Civil Engineers (ASCE). 2009. *Failing Infrastructure Cannot Support a Healthy Economy.* Washington, D.C.: American Society of Civil Engineers.

Archer, C. L., and M. Z. Jacobson. 2003. The Spatial and Temporal Distributions of U.S. Winds and Wind Power at 80 m Derived from Measurements. *Journal of Geophysical Research* 108 (D9), doi:10.1029/2002JD002076.

———. 2005. Evaluation of Global Wind Power. *Journal of Geophysical Research* 110, doi:10.1029/2004JD005462.

Atkins, S. E. 2000. *Historical Encyclopedia of Atomic Energy.* Westport, Conn.: Greenwood Press.

Awerbuch, S. 2006. Wind Economics in the 21st Century. *Wind Directions.* January/February, 44–47.

Beer, C., W. Lucht, C. Schmullius, and A. Shvidenko. 2006. Small Net Carbon Dioxide Uptake by Russian Forests during 1981–1999. *Geophysical Research Letters* 33 (15): L15403.

Bellamy, P. H., P. J. Loveland, R. I. Bradley, R. M. Lark, and G. J. D. Kirk. 2005. Carbon Losses from All Soils across England and Wales 1978–2003. *Nature* 437:245–48.

Bennett, R. K. 2008. Why Gasoline Is Still King. *The American.* December 17. http://www.american.com/archive/2008/november-december-magazine/why-gasoline-is-still-king (accessed January 5, 2010).

Berdichevsky, G., K. Kelty, J. B. Straubel, and E. Toomre. 2006. *The Tesla Roadster Battery System.* http://www.teslamotors.com/display_data/TeslaRoadsterBatterySystem.pdf (accessed January 5, 2010).

Bethe, H. 1977. The need for nuclear power. *Bulletin of the Atomic Scientists* 33 (3):59–63.

Better Place. 2009. The Solution. http://www.betterplace.com/solution (accessed January 5, 2010).

Blanco-Canqui, H., R. Lal, W. M. Post, R. C. Izaurralde, and M. J. Shipitalo. 2007. Soil Hydraulic Properties Influenced by Corn Stover Removal from No-Till Corn in Ohio. *Soil and Tillage Research* 92:144–55.

Boccard, N. 2009. Capacity Factor of Wind Power Realized Values vs. Estimates. *Energy* 37:2679–88.

Brandt, A. R. 2007. Testing Hubbert. *Energy Policy* 35:3074–88.

Brazil. Government of the State of São Paulo. Secretariat of the Environment. 2004. *Assessment of Greenhouse Gas Emissions in the Production and Use of Fuel Ethanol in Brazil.* By I. C. Macedo, M. R. L. V. Leal, and J. E. A. R. da Silva, University of Campinas (UNICAMP). http://www.wilsoncenter.org/events/docs/brazil.unicamp.macedo.greenhousegas.pdf (accessed December 16, 2009).

———. Ministério de Agricultura, Pecuária e Abastecimento. 2007. *Ethanol: The Brazilian Experience.* By A. Bressan and E. Contini.

Briggs, M. 2009. Online comment in response to "A Plan to Power 100 Percent of the Planet with Renewables," by M. Z. Jacobson and M. A. Delucchi, in the November 2009 issue of *Scientific American.* October 26, 2009. http://www.scientificamerican.com/article.cfm?id=a-path-to-sustainable-energy-by-2030 (accessed January 5, 2010).

British Petroleum (BP). Various years. *BP Statistical Review of World Energy.* Report for 2009 is available at http://www.bp.com/liveassets/bp_internet/globalbp/globalbp_uk_english/reports_and_publications/statistical_energy_review_2008/STAGING/local_assets/2009_downloads/statistical_review_of_world_energy_full_report_2009.pdf (accessed January 14, 2010). Historical data for 1965–2008 are available at the same website.

British Wind Energy Association (BWEA). 2005. *Wind Turbine Technology.* London: British Wind Energy Association. http://www.bwea.com/pdf/briefings/technology05_small.pdf (accessed January 5, 2010).

Brown, L. R. 2003. Wind Power Set to Become World's Leading Energy Source. Earth Policy Institute. June 25. http://www.earth-policy.org/Updates/Update24.htm (accessed January 5, 2010).

Brumfiel, G. 2008. Nuclear Renaissance Plans Hit by Financial Crisis. *Nature* 456:286–87.

Bunker, J. G. 1972. *Liberty Ships.* New York: Arno Press.

Bupp, I. C., and J. C. Derian. 1978. *Light Water: How the Nuclear Dream Dissolved.* New York: Basic Books.

Burwell, C. C. 1990. Transportation: Electricity's Changing Importance over Time. In *Electricity in the American Economy*, ed. S. H. Schurr, C. C. Burwell, W. D. Devine Jr., and S. Sonenblum, 209–31. New York: Greenwood Press.

Cairns, E. J. 2004. Battery Overview. In *Encyclopedia of Energy*, vol. 1, ed. C. J. Cleveland, 124. New York: Elsevier.

Cambridge Energy Research Associates (CERA). 2006. *Why the "Peak Oil" Theory Falls Down—Myths, Legends, and the Future of Oil Resources*. Cambridge, Mass.: Cambridge Energy Research Associates.

Campbell, J. E., D. B. Lobell, and C. B. Field. 2009. Greater Transportation Energy and GHG Offsets from Bioelectricity than Ethanol. *Science* 324:1055–57.

Canadian Association of Petroleum Producers (CAPP). 2009. *Statistical Handbook*. http://www.capp.ca/library/statistics/handbook/Pages/default.aspx (accessed January 5, 2010).

Carey, E. V., A. Sala, R. Keane, and R. M. Callaway. 2001. Are Old Forests Underestimated as Global Carbon Sinks? *Global Change Biology* 7:339–44.

Carin, R. 1969. *Power Industry in Communist China*. Hong Kong: Union Research Institute.

Carter, J. 1979. The "Crisis of Confidence" speech, televised on July 15, 1979. http://www.pbs.org/wgbh/amex/carter/filmmore/ps_crisis.html (accessed January 5, 2010).

Cassmann, K. G., A. Dobermann, and D. T. Walters. 2002. Agroecosystems, Nitrogen-Use Efficiency, and Nitrogen Management. *Ambio* 31:132–40.

Cavallo, A. J. 1995. High-Capacity Factor Wind Energy Systems. *Journal of Solar Energy Engineering* 117:137–43.

Central Intelligence Agency (CIA). 1979. *The World Oil Market in the Years Ahead*. Washington, D.C.: Central Intelligence Agency.

Chandler, G. 2007. Weyburn Project Sets CO_2 Sequestration on World Stage. *Alberta Oil* 3 (1): 28–31.

Chapin, D. M., K. P. Cohen, W. K. Davis, E. E. Kintner, L. J. Koch, J. W. Landis, M. Levenson, et al. 2003. Nuclear Power Plants and Their Fuel as Terrorist Targets. *Science* 297:1997–99.

Charles, D. 2007. Spinning a Nuclear Comeback. *Science* 315:1782–94.

Chevalier, R. 2000. Hypercar! The People's Car! http://www.remyc.com/hypercar.html (accessed January 5, 2010).

China Daily. 2009. China Nuclear Power Installed Capacity May Top 70 GW by 2020. http://www.chinadaily.com.cn/bizchina/2009-11/03/content_8905754.htm (accessed January 5, 2010).

Chu, S. 2009. Carbon Capture and Sequestration. *Science* 325:1599.

Citizens against CO_2 Sequestration. 2009. http://citizensagainstco2sequestration.blogspot.com (accessed January 5, 2010).

Coale, K. H., K. S. Johnson, F. P. Chavez, K. O. Buesseler, R. T. Barber, M. A. Brzezinski, W. P. Cochlan, et al. 2004. Southern Ocean Iron Enrichment Experiment: Carbon Cycling in High- and Low-Si Waters. *Science* 304:408–14.

Cohn, S. M. 1997. *Too Cheap to Meter: An Economic and Philosophical Analysis of the Nuclear Dream*. Albany, N.Y.: State University of New York Press.

Commuter Cars Corporation. n.d. Commuter Cars: Tango. http://www.commutercars.com/home.html (accessed January 4, 2010).

Cowan, R. 1990. Nuclear Power Reactors: A Study in Technological Lock-In. *Journal of Economic History* 50:541–67.

Creagan, R. J. 1973. Boon to Society: The LMFBR. *Power Engineering* 77 (2): 12–16.

Criswell, D. 2000. Lunar Solar Power System: Review of the Technology Base of an Operational LSP System. *Acta Astronautica* 46:531–40.

Daimler. 2009. DiesOtto—Gasoline Engine with the Diesel Genes. http://www.daimler.com/dccom/0-5-962545-1-962547-1-0-0-0-0-0-36-7165-0-0-0-0-0-0-0. html (accessed January 5, 2010).

Dayal, S. 2009. Online comment in response to "A Plan to Power 100 Percent of the Planet with Renewables," by M. Z. Jacobson and M. A. Delucchi, in the November 2009 issue of *Scientific American*. October 26, 2009. http://www.scientificamerican.com/article.cfm?id=a-path-to-sustainable-energy-by-2030 (accessed January 15, 2010).

De Oliveira, M. E. D., B. E. Vaughan, and E. J. Rykiel. 2005. Ethanol as Fuel: Energy, Carbon Dioxide Balances, and Ecological Footprint. *BioScience* 55:593–602.

De Vries, W., G. J. Reinds, P. Gundersen, and H. Sterba. 2006. The Impact of Nitrogen Deposition on Carbon Sequestration in European Forests and Forest Soils. *Global Change Biology* 12:1151–73.

DeCarolis, J. F., and D. Keith. 2001. The Real Cost of Wind Energy. *Science* 294:1000–1001.

Deffeyes, K. S. 2004. Current Events: Join Us as We Watch the Crisis Unfolding. January 16. http://www.princeton.edu/hubbert/current-events-04-01.html (accessed January 5, 2010).

Denning, R. S. 1985. The Three Mile Island Unit's Core: A Post-Mortem Examination. *Annual Review of Energy* 10:35–52.

Deutsche Energie-Agentur (DENA). 2005. *Planning of the Grid Integration of Wind Energy in Germany Onshore and Offshore Up to the Year 2020*. Berlin: Deutsche Energie-Agentur.

Deutsche Welle. 2009. Germany Plans to Put One Million Electric Cars on the Road by 2020. August 19. http://www.dw-world.de/dw/article/0,,4582176,00. html (accessed January 5, 2010).

Domrös, M., and G. Peng. 1988. *The Climate of China*. Berlin: Springer-Verlag.

Dooley, J. J., R. T. Dahowski, and C. L. Davidson. 2009. Comparing Existing Pipeline Networks with the Potential Scale of Future U.S. CO_2 Pipeline Networks. In *Energy Procedia: 9th International Conference on Greenhouse Gas Control Technologies* (GHGT9) 1 (1): 1595–1602. Elsevier, London, United Kingdom. doi:10.1016/j.egypro.2009.01.209.

Duncan, R. C. 1998. The Olduvai Theory: Sliding Towards a Post-Industrial Stone Age. http://dieoff.org/page125.htm (accessed January 4, 2010).

———. 2000. The Peak of World Oil Production and the Road to the Olduvai Gorge. Paper presented at the Pardee Keynote Symposia, Geological Society of America Summit, Reno, Nevada, November 13.

Durante, D., and M. Miltenberger. 2004. *Net Energy Balance of Ethanol Production.* Sioux Falls, South Dakota. American Coalition for Ethanol. http://www.ethanol.org/pdf/contentmgmt/Issue_Brief_Ethanols_Energy_Balance.pdf (accessed January 5, 2010).

Eberhard, M., and M. Tarpenning. 2006. *The 21st Century Electric Car.* San Carlos, Calif.: Tesla Motors. http://energy.wesrch.com/Paper/paper_details.php?id= TR1SVV3O7X124&paper_type=pdf&type=%20viewing (accessed January 14, 2010).

El-Badri, A. S. 2008. Oil Markets Today and Challenges Ahead. Speech delivered at the Third International Energy Week, Moscow, October 22–24. http://www.opec.org/opecna/speeches/2008/SGMoscow.htm (accessed January 5, 2010).

Electrifying Times. 2007. Clooney's Tango! WoW!!! http://www.electrifyingtimes.com/ClooneyTango.html (accessed January 5, 2010).

Elvidge, C. D. 2004. U.S. Constructed Area Approaches the Size of Ohio. *Eos* 85:233–34.

ENERCON. 2009. E-82. http://www.enercon.de/en/_home.htm (accessed January 6, 2010).

European Federation for Transport and Environment (EFTE). 2009. *How to Avoid Electric Shock. Electric Cars: from Hype to Reality.* Brussels: Transport and Environment. http://www.transportenvironment.org/Publications/prep_hand_out/lid:560 (accessed January 6, 2010).

European Wind Energy Association (EWEA). 2007. *The Myth of Intermittency.* http://www.ewea.org/index.php?id=242&no_cache=1&sword_list[]=intermittency (accessed January 6, 2010).

——— and Greenpeace. 2004. *Wind Force 12: A Blueprint to Achieve 12% of the World's Electricity from Wind Power by 2020.* Brussels: EWEA. http://www.ewea.org/fileadmin/ewea_documents/documents/publications/WF12/WF12-2004_eng.pdf (accessed January 6, 2010).

EV Innovations Inc. (EVII). 2009. Products. http://www.hybridtechnologies.com/products.php (accessed January 4, 2010).

Fairless, D. 2007. The Little Shrub That Could—Maybe. *Nature* 449:652–55.

Fan, S., M. Gloor, J. Mahlman, S. Pacala, J. Sarmiento, T. Takahashi, and P. Tans. 1998. A Large Terrestrial Carbon Sink in North America Implied by Atmospheric and Oceanic Carbon Dioxide Data and Models. *Science* 282:442–46.

Farah, A. 2008. Chevy Volt Update from Chief Engineer: "There's Nothing Standing in Our Way." August 13. http://gm-volt.com/2008/08/13/chevy-volt-update-from-chief-engineer-nothing-standing-in-our-way (accessed January 6, 2010).

Feldman, S. L., M. A. Bernstein, and R. B. Noland. 1988. The Costs of Completing Unfinished U.S. Nuclear Power Plants. *Energy Policy* 16:270–79.

Felix, E., and D. R. Tilley. 2009. Integrated Energy, Environmental and Financial Analysis of Ethanol Production from Cellulosic Switchgrass. *Energy* 34:410–36.

Ferguson, C. D., P. D. Reed, and M. M. Smith. 2009. The Nuclear Option. *Foreign Policy* January/February: 40–41.

Ferrett, G. 2007. Biofuel's "Crime against Humanity." *BBC Online*. October 27. http://news.bbc.co.uk/1/hi/world/americas/7065061.stm (accessed January 6, 2010).

Flower, A. 1978. World Oil Production. *Scientific American* 238 (3): 42–49.

Ford, Henry. 1922. *My Life and Work*. New York: Doubleday.

Ford Motor Company (FMC). 1909. *Ford Motor Cars*. Detroit, Mich.: FMC. http://www.mtfca.com (accessed January 6, 2010).

Garwin, R. L., and G. Charpak. 2001. *Megawatts and Megatons: The Future of Nuclear Power and Nuclear Weapons*. Chicago: University of Chicago Press.

General Motors. 2009. Chevy Volt FAQs. http://gm-volt.com/chevy-volt-faqs (accessed January 6, 2010).

Giampietro, M., and K. Mayumi. 2009. *The Biofuel Delusion: The Fallacy of Large-Scale Agro-Biofuel Production*. London: Earthscan.

Gibson, D. G., G. A. Benders, C. Andrews-Pfannkoch, E. A. Denisova, H. Baden-Tillson, J. Zaveri, T. B. Stockwell, et al. 2008. Assembly and Cloning of a *Mycoplasma Genitalium* Genome. *Science* 319:1215–20.

Gilfillan, S. M. V., B. S. Lollar, G. Holland, D. Blagburn, S. Stevens, M. Schoell, M. Cassidy, Z. Ding, Z. Zhou, G. Lacrampe-Couloume, and C. J. Ballentine. 2009. Solubility Trapping in Formation Water as Dominant CO_2 Sink in Natural Gas Fields. *Nature* 458:614–18.

Gillis, J. 2006. Stuck in Neutral: America's Failure to Improve Motor Vehicle Fuel Efficiency 1996–2005. Washington, D.C.: Consumer Federation of America. http://www.consumerfed.org/pdfs/Stuck_in_Neutral.pdf (accessed January 6, 2010).

Global Wind Energy Council (GWEC). 2009. Global Installed Wind Power Capacity (MW)—Regional Distribution. http://www.gwec.net/fileadmin/documents/PressReleases/PR_stats_annex_table_2nd_feb_final_final.pdf (accessed January 6, 2010).

Goldberg, D. S., T. Takahashi, and A. L. Slagle. 2008. Carbon Dioxide Sequestration in Deep-Sea Basalt. *Proceedings of the National Academy of Sciences* 105:9920–25.

Gore, A. 2008. A Generational Challenge to Repower America. http://www.algore.org/generational_challenge_repower_america_al_gore (accessed January 6, 2010).

Government of Alberta. 2010. Oil sands. http://www.energy.alberta.ca/OurBusiness/oilsands.asp.

Graham, R. L., R. Nelson, J. Sheehan, R. D. Perlack, and L. L. Wright. 2007. Current and Potential U.S. Corn Stover Supplies. *Agronomy Journal* 99:1–11.

Greenpeace International. 2008. *False Hope: Why Carbon Capture and Storage Won't Save the Climate*. Amsterdam: Greenpeace International. http://www.greenpeace. org/raw/content/international/press/reports/false-hope.pdf (accessed January 6, 2010).

———— and European Renewable Energy Council (EREC). 2007. *Energy (R)evolution*. Amsterdam: Greenpeace International/ European Renewable Energy Council.

Greenspun, P. 2008. Cost of Converting Entire U.S. to Electric Cars? Zero. May 27. http://blogs.law.harvard.edu/philg/2008/05/27/cost-of-converting-entire-us-to-electric-cars-zero (accessed January 6, 2010).

Guru, M. V., and J. E. Horn. 2000. *The Ogallala Aquifer*. Poteau, Okla.: Kerr Center for Sustainable Agriculture. http://www.kerrcenter.com/publications/ogallala_ aquifer.pdf (accessed January 6, 2010).

Gustavson, M. R. 1979. Limits to Wind Power Utilization. *Science* 204:13–17.

Hampton, N., R. Hartlein, H. Lennartsson, H. Orton, and R. Ramachandran. 2007. Long-Life XLPE Insulated Power Cable. http://www.neetrac.gatech.edu/ publications/jicable07_C_5_1_5.pdf (accessed January 6, 2010).

Hawken, P. 1997. Hypercar. *Mother Jones*. March/April. http://www.motherjones. com/news/feature/1997/03/hawken7.html (accessed January 6, 2010).

Hayes, D., H. Epstein, S. Lawrence, J. Dierker, T. Cohen, G. Thompson, P. Duel, G. Deloss, J. Lash, and J. Gibson.1979. *Blueprint for a Solar America*. Washington, D.C.: Solar Lobby.

Heath, J., E. Ayres, M. Possell, R. D. Bardgett, H. I. J. Black, H. Grant, H., P. Ineson, and G. Kerstiens. 2005. Rising Atmospheric CO_2 Reduces Sequestration of Root-Derived Soil Carbon. *Science* 309:1711–13.

Herring, H. 2006. Energy Efficiency—A Critical View. *Energy* 31:10–20.

Hilyard, J., ed. 2008. *2008 International Petroleum Encyclopedia*. Tulsa, Okla.: PennWell Books.

Himmel, M. E., T. Vinzant, S. Bower, and J. Jechura. 2007. Biomass Recalcitrance: Engineering Plants and Enzymes for Biofuels Production. *Science* 315:804–7.

Hohenemser, C. 1988. The Accident at Chernobyl: Health and Environmental Consequences and the Implications for Risk Management. *Annual Review of Energy* 13:383–428.

————, R. Kasperson, and R. Kates. 1977. The Distrust of Nuclear Power. *Science* 196:25–34.

Howarth, R. W., and S. Bringezu, eds. 2009. *Biofuels: Environmental Consequences and Interactions with Changing Land Use*. Ithaca, New York: Cornell University.

Huang, Z., and Z. Chang. 1980. Development of Methane Is an Important Task in Solving the Rural Energy Problem. *Hongqi* (Red Flag) 21:39–41.

Hubbert, M. K. 1956. Nuclear Energy and Fossil Fuels. In *Drilling and Production Practice*, 7–25. Washington, D.C.: American Petroleum Institute.

————. 1969. Energy Resources. In *Resources and Man*, ed. National Academy of Sciences–National Research Council, 157–242. San Francisco: W. H. Freeman.

IHS Global Insight. 2009. World Car Industry Forecast Service. http://www.
ihsglobalinsight.com/ProductsServices/ProductDetail727.htm (accessed January 6, 2010).

Imbrecht, C. R. 1995. California's Electrifying Future. Written for *Electric and Hybrid Vehicle Technology 95 Magazine* for distribution at Second ITS World Conference, Yokohama, Japan, November 1995, and ENV, Detroit, Michigan, USA, January 1996. http://www.energy.ca.gov/papers/CEC-999-1995-002.txt (accessed January 6, 2010).

Informationsdienst Wissenschaft (IDW). 2009. LOHAFEX Provides New Insights on Plankton Ecology. http://idw-online.de/pages/de/news306656 (accessed January 10, 2010).

Intel. 2007. *60 Years of the Transistor: 1947–2007*. http://www.intel.com/technology/timeline.pdf (accessed January 6, 2010).

———. 2010. *Moore's Law*. http://www.intel.com/technology/mooreslaw/ (accessed January 10, 2010).

Intergovernmental Panel on Climate Change (IPCC). 2006. *Special Report on Carbon Dioxide Capture and Storage*. Geneva: Intergovernmental Panel on Climate Change.

International Atomic Energy Agency (IAEA). 2001. *Sustainable Development and Nuclear Power*. Vienna: International Atomic Energy Agency.

———. 2006. Status of Innovative Small and Medium Sized Reactor Designs. 2005: Reactors with Conventional Refueling Schemes. Vienna: International Atomic Energy Agency.

———. 2008. Latest News Related to PRIS and the Status of Nuclear Power Plants. http://www.iaea.org/cgi-bin/db.page.pl/pris.main.htm (accessed January 6, 2010).

International Commission on Large Dams (ICOLD). 1998. *World Register of Dams*. Paris: International Commission on Large Dams.

International Energy Agency (IEA). 2005. *Energy Technologies at the Cutting Edge*. Paris: International Energy Agency.

———. 2006. CO_2 *Capture and Storage*. Paris: International Energy Agency.

———. 2008. *Outlook for Hybrid and Electric Vehicles*. http://www.ieahev.org/pdfs/ia-hev_outlook_2008.pdf (accessed January 6, 2010).

———. 2009. *Renewables Information*. Paris: International Energy Agency.

InterTechnology Corporation. 1977. *An Analysis of the Economic Potential of Solar Thermal Energy to Provide Industrial Process Heat*. Warrenton, Va.: InterTechnology Corporation.

Ivanhoe, L. F. 1995. Future World Oil Supplies: There Is a Finite Limit. *World Oil* 216 (10): 77–79.

Jacobson, M. Z., and M. A. Delucchi. 2009. A Plan to Power 100 Percent of the Planet with Renewables. *Scientific American* 301 (5): 58–65.

Jacobson, M. Z., and G. M. Masters. 2001a. Exploiting Wind versus Coal. *Science* 293:1348.

———. 2001b. The Real Cost of Wind Energy. *Science* 294:1001–2.

Jaffe, A. M. 2007. Flip the Switch. *Foreign Policy* May/June, 44.

Janssens, I. A., A. Freibauer, B. Schlamadinger, R. Ceulemans, P. Ciais, A. J. Dolman, M. Heimann, et al. 2005. The Carbon Budget of Terrestrial Ecosystems at Country-Scale—A European Case Study. *Biogeosciences* 2:15–26.

Japan Nuclear Cycle Development Institute (JNCDI). 2000. *The Monju Sodium Leak*. Tokyo: Japan Nuclear Cycle Development Institute.

Jayaraman, K. S. 2007. India's Carbon Dioxide Trap. *Nature* 445:350.

Jevons, W. S. 1865. *The Coal Question: An Inquiry Concerning the Progress of the Nation, and the Probable Exhaustion of Our Coal Mines*. London: Macmillan.

Jochem, E., ed. 2002. *Steps Towards a Sustainable Development*. Zurich: CEPE/ETH and Novatlantis. http://www.novatlantis.ch/fileadmin/downloads/2000watt/Weissbuch.pdf (accessed January 6, 2010).

Johansson, T., and P. Steen. 1978. *Solar Sweden: An Outline of a Renewable Energy System*. Stockholm: Secretariat for Futures Studies.

Josephson, M. 1959. *Edison: A Biography*. New York: McGraw Hill.

Kadam, K. L., and J. D. McMillan. 2003. Availability of Corn Stover as a Sustainable Feedstock for Bioethanol Production. *Bioresource Technology* 88:17–25.

Keith, D. W. 2009. Why Capture CO_2 from the Atmosphere? *Science* 325:1654–55.

———, and A. Adams. 2009. Climatic Limits to Wind Capacity. Paper in preparation. University of Calgary, Calgary, AB.

Keith, D. W., J. F. DeCarolis, D. C. Denkenberger, D. H. Lenschow, S. L. Malyshev, S. Pacala, and P. J. Rasch. 2004. The Influence of Large-Scale Wind Power on Global Climate. *Proceedings of the National Academy of Sciences* 101:16115–20.

Kelemen, P. B., and J. Matter. 2008. In Situ Carbonation of Peridotite for CO_2 Storage. *Proceedings of the National Academy of Sciences* 105:17295–300.

Kerr, R. A. 2007. The Looming Oil Crisis Could Arrive Uncomfortably Soon. *Science* 316:351.

Kharaka, Y. K., D. R. Cole, S. D. Hovorka, W. D. Gunter, W. D. Knauss, K. G. Knauss, and B. M. Freifeld. 2006. Gas–Water–Rock Interactions in Frio Formation Following CO_2 Injection: Implications for the Storage of Greenhouse Gases in Sedimentary Basins. *Geology* 34:577–80.

Khosla, V. 2006. A Healthier Addiction. *Economist*. March 25, 72.

Kim, S., and B. E. Dale. 2002. Allocation Procedure in Ethanol Production System from Corn Grain. *International Journal of Life Cycle Assessment* 7 (4): 237–43.

Kintner-Meyer, M., K. Schneider, and R. Pratt. 2007. Impacts Assessment of Plug-In Hybrid Vehicles on Electric Utilities and Regional Power Grids. Part I: Technical Analysis. *Journal of EUEC* 1: paper no. 04. http://www.euec.com/documents/pdf/Paper_4.pdf (accessed December 16, 2009).

Kirsch, D. 2000. *The Electric Vehicle and the Burden of History*. New Brunswick, N.J.: Rutgers University Press.

Kondamudi, N., S. K. Mohapatra, and M. Misra. 2008. Spent Coffee Grounds as a Versatile Source of Green Energy. *Journal of Agricultural and Food Chemistry* 56:11757–60.

Koplow, D. 2006. *Biofuels—At What Cost? Government Support for Ethanol and Biodiesel in the United States.* Geneva: International Institute for Sustainable Development.

Kurzweil, R., and C. Meyer. 2003. Understanding the Accelerating Rate of Change. http://www.kurzweilai.net/meme/frame.html?main=/articles/art0563.html (accessed January 6, 2010).

Kwok, R. 2009. Cellulosic Ethanol Hits Roadblocks. *Nature* 461:582–83.

Lackner, K. S., H. J. Ziock, and P. Grimes. 1999. The Case for Carbon Dioxide Extraction from Air. *SourceBook* 57:6–10.

Lal, R., M.Griffin, J. Apt, L. Lave, and M. G. Morgan. 2004. Managing Soil Carbon. *Science* 304:393.

Lazaroff, Cat. 2001. California Mandates Electric Cars. *Environment News Service.* January 30. http://www.mail-archive.com/sustainablelorgbiofuel@sustainablelists.org/msg02490.html (accessed January 6, 2010).

Lehmann, J. 2007. A Handful of Carbon. *Nature* 447:143–44.

———, J. Gaunt, and M. Rondon. 2006. Bio-Char Sequestration in Terrestrial Ecosystems—A Review. *Mitigation and Adaptation Strategies for Global Change* 11:403–27.

Lilienthal, D. E. 1959. *The Journals of David E. Lilienthal: The Road to Change, 1955–1959.* New York: Harper and Row.

Liu, E. L. 2006. Imperial Oil—A Leader in Thermal In-Situ Production. Speech before the Edmonton Society of Financial Analysts. June 8. http://www.imperialoil.ca/Canada-English/files/News/N_S_Speech060608.pdf (accessed January 6, 2010).

Lorenz, E. N. 1976. *The Nature and Theory of the General Circulation of the Atmosphere.* Geneva: World Meteorological Organization.

Lovelock, J. 2006. *The Revenge of Gaia: Why the Earth Is Fighting Back—And How We Can Still Save Humanity.* London: Allen Lane.

Lovins, A. B. 1976. Energy Strategy: The Road Not Taken. *Foreign Affairs* 55 (1): 65–96.

———. 1978. Soft Energy Technologies. *Annual Review of Energy* 3:477–517.

———. 1992. The Soft Path—Fifteen Years Later. *Rocky Mountain Institute Newsletter* 8 (1): 9.

Lowen, R. S. 1987. Entering the Atomic Power Race: Science, Industry, and Government. *Political Science Quarterly* 102:459–79.

Loya, W. M., K. S. Pregitzer, N. J. Karberg, J. S. King, and C. P. Giardina. 2003. Reduction of Soil Carbon Formation by Tropospheric Ozone under Increased Carbon Dioxide Levels. *Nature* 425:705–6.

Lu, X., M. B. McElroy, and J. Kiviluoma. 2009. Global Potential for Wind-Generated Electricity. *Proceedings of the National Academy of Sciences* 106:10933–38.

Lugar, R. G., and R. J. Woolsey. 1999. The New Petroleum. *Foreign Affairs* 78 (1): 88–102.

Magenn Power. 2009. Magenn Power Air Rotor Products. http://www.magenn.com/products.php (accessed January 6, 2010).

Makhijani, A., and S. Saleska. 1999. *The Nuclear Power Deception: U.S. Nuclear Mythology from Electricity "Too Cheap to Meter" to "Inherently Safe" Reactors*. New York: Apex Press.

Marchetti, C. 1989. How to Solve the CO_2 Problem without Tears. *International Journal of Hydrogen Energy* 14:493–506.

Marland, G., T. Boden, and R. J. Andres. 2007. *Global CO_2 Emissions from Fossil-Fuel Burning, Cement Manufacture, and Gas Flaring: 1751–2004*. Oak Ridge, Tenn.: Oak Ridge National Laboratory.

Marris, E. 2006. Nuclear Reincarnation. *Nature* 441:796–97.

———, and D. Fairless. 2007. Wind Farms' Deadly Reputation Hard to Shift. *Nature* 447:126.

Marshall, E. 2005. Is the Friendly Atom Poised for a Comeback? *Science* 309:1168–69.

Martin, J. H., and S. E. Fitzwater. 1988. Iron-Deficiency Limits Phytoplankton Growth in the Northeast Pacific Subarctic. *Nature* 331:341–43.

Mascarelli, A. L. 2009. Gold Rush for Algae. *Nature* 461:460–61.

McGowan, J., and Connors S. 2000. Windpower—A Turn of the Century Review. *Annual Review of Energy and the Environment* 25:147–97.

McShane, Clay. 1997. *The Automobile: A Chronology*. New York: Greenwood Press.

Meier, R. L. 1956. *Science and Economic Development*. Cambridge, Mass.: MIT Press.

Moore, G. E. 1965. Cramming More Components onto Integrated Circuits. *Electronics* 38 (8): 114–17.

Moore, P. 2006. Going Nuclear. *Washington Post*. April 16.

Murphy, P. M. 1974. *Incentives for the Development of the Fast Breeder Reactor*. Stamford, Conn.: General Electric.

Nabuurs, G. J., M. J. Schelhaas, G. M. J. Mohren, and C. B. Field. 2003. Temporal Evolution of the European Forest Carbon Sink from 1950 to 1999. *Global Change Biology* 9:152–60.

National Renewable Energy Laboratory (NREL). 2003. National Renewable Energy Laboratory 2003 Research Review. http://www.nrel.gov/docs/fy04osti/36178.pdf (accessed January 10, 2010).

———. 2009. Best Research-Cell Efficiencies. http://www.nrel.gov/pv/thin_film/docs/kaz_best_research_cells.ppt (accessed January 10, 2010).

Nazri, G. A., and G. Pistoia, eds. 2004. *Lithium Batteries: Science and Technology*. Boston: Kluwer Academic.

Nehring, R. 2006a. Two Basins Show Hubbert's Method Underestimates Future Oil Production. *Oil and Gas Journal* 104 (13): 37–44.

———. 2006b. How Hubbert's Method Fails to Predict Oil Production in the Permian Basin. *Oil and Gas Journal* 104 (15): 30–35.

———. 2006c. Post-Hubbert Challenge Is to Find New Methods to Predict Production, EUR. *Oil and Gas Journal* 104 (16): 43–51.

Nemani, R., M. White, P. Thornton, K. Nishida, S. Reddy, J. Jenkins, and S. Running. 2002. Recent Trends in Hydrologic Balance Have Enhanced the Terrestrial Carbon Sink in the United States. *Geophysical Research Letters* 29:106-1–106-4.

Nixon, R. M. 1974. State of the Union Address 1974. http://stateoftheunionaddress.org/category/richard-nixon (accessed January 6, 2010)

Nord Pool. 2009. No.16/2009 Nord Pool Spot Implements Negative Price Floor in Elspot from October 2009. http://www.nordpoolspot.com/Market_Information/Exchange-information/No162009-Nord-Pool-Spot-implements-negative-price-floor-in-Elspot-from-October-2009- (accessed January 6, 2010).

Nuclear Energy Agency (NEA). 2002. *Chernobyl: Assessment of Radiological and Health Impacts*. Paris: Nuclear Energy Agency.

Nuclear Information and Resource Service (NIRS). 1999. Background on Nuclear Power and Kyoto Protocol. http://www.nirs.org/globalization/CDM-Nukesnirsbackground.htm (accessed January 6, 2010).

Oak Ridge National Laboratory (ORNL). 1968. *Nuclear Energy Centers: Industrial and Agro-Industrial Complexes*. Oak Ridge, Tenn.: Oak Ridge National Laboratory.

Odell, P. R. 2006. The Response by Peter R. Odell, Professor Emeritus of International Energy Studies, Erasmus University Rotterdam, on the occasion of his acceptance of the Biennial OPEC Award. Third OPEC International Seminar, Vienna, Austria, September 12–15. http://www.gasresources.net/Odell433.09%20-%20(3rd%20OPEC%20International%20Seminar,%20Vienna).pdf (accessed January 6, 2010).

Olah, G. A., A. Goeppert, and G. K. S. Prakash. 2006. *Beyond Oil and Gas: The Methanol Economy*. Weinheim, Germany: Wiley-VCH.

Olds, F. C. 1982. A New Look at Nuclear Power Costs. *Power Engineering* 86 (2): 34–42.

Omega Research. 1997. TFC Commodity Charts. Light Crude Oil (CL, NYMEX) Weekly Price Chart. http://futures.tradingcharts.com/chart/CO/W (accessed January 4, 2010).

Organisation for Economic Cooperation and Development (OECD). 2000. *Nuclear Energy in a Sustainable Development Perspective*. Paris: Organisation for Economic Cooperation and Development.

Organization of the Petroleum Exporting Countries. 2009. Reference Prices: OPEC Basket Price. http://www.opec.org/home/basket.aspx (accessed January 4, 2010).

Østergaard, P. A. 2008. Geographic Aggregation and Wind Power Output Variance in Denmark. *Energy* 33:1453–60.

OTEC News. 2009. What is OTEC? http://www.otecnews.org/whatisotec.html (accessed December 2, 2009).

Pacala, S., and R. Socolow. 2004. Stabilization Wedges: Solving the Climate Problem for the Next 50 Years with Current Technologies. *Science* 305:968–72.

Paine, C. 2006. *Who Killed the Electric Car?* New York: Sony Pictures Classics. http://www.sonyclassics.com/whokilledtheelectriccar (accessed January 6, 2010).

Parkins, W. E. 2006. Fusion Power: Will It Ever Come? *Science* 311:1380.

Parrish, D. J., and J. H. Fike. 2005. The Biology and Agronomy of Switchgrass for Biofuels. *Critical Reviews in Plant Sciences* 24:423–59.

Pasqualetti, M., P. Gipe, and R. W. Righter, eds. 2002. *Wind Power in View*. San Diego, Calif.: Academic Press.

Patzek, T. W., and D. Pimentel. 2005. Thermodynamics of Energy Production from Biomass. *Critical Reviews in Plant Sciences* 24:327–64.

Peixoto, J. P., and A. H. Oort. 1992. *Physics of Climate*. New York: American Institute of Physics.

Percy, K. E., R. Jandl, J. P. Hall, and M. Lavigne. 2009. *The Role of Forests in Carbon Cycles, Sequestration, and Storage*. Vienna: IUFRO. http://www.iufro.org/download/file/1626/3754/issue1_march31.pdf (accessed January 6, 2010)

Perlack, R. D., L. L. Wright, A. F. Turhollow, R. L. Graham, B. J. Stokes, and D. C. Erbach. 2005. *Biomass as Feedstock for a Bioenergy and Bioproducts Industry: The Technical Feasibility of a Billion-Ton Annual Supply*. Oak Ridge, Tenn.: Oak Ridge National Laboratory.

Pickens, T. B. 2008. T. Boone Pickens Unveils *The Pickens Plan*—A Sweeping, Innovative Plan to Address National Energy Dependency Crisis. Press release. July 8. http://media.pickensplan.com/presskit/2008/pickensplan_press_unveils.pdf (accessed December 2, 2009).

Pimentel, D. 2003. Ethanol Fuels: Energy Balance, Economics, and Environmental Impacts Are Negative. *Natural Resources Research* 12:127–34.

———, and T. W. Patzek. 2005. Ethanol Production Using Corn, Switchgrass, and Wood: Biodiesel Production Using Soybean and Sunflower. *Natural Resources Research* 14:65–76.

Polimeni, J., K. Mayumi, and M. Giampietro. 2007. Jevons' Paradox: The Myth of Resource Efficiency Improvements. London: Earthscan.

Pordesimo, L. O., W. C. Edens, and S. Sokhansanj. 2004. Distribution of Aboveground Biomass in Corn Stover. *Biomass and Bioenergy* 26:337–43.

Potter, C., S. Klooster, R. Myneni, V. Genovese, P. Tan, and V. Kumar. 2003. Continental Scale Comparisons of Terrestrial Carbon Sinks Estimated from Satellite Data and Ecosystem Modeling, 1982–89. *Global and Planetary Change* 39:201–3.

Power. 2009. T. Boone Pickens Suspends Mega-Wind Farm in Texas. July 8. http://www.powermag.com/POWERnews/T-Boone-Pickens-Suspends-Mega-Wind-Farm-in-Texas_2037.html (accessed January 6, 2010).

Power-Technology.com. 2009. Bavaria Solarpark, the World's Largest Project, Bavaria, Germany. http://www.power-technology.com/projects/bavaria/ (accessed December 3, 2009).

Rabalais, N. N. 2002. Nitrogen in Aquatic Ecosystems. *Ambio* 31:102–12.

Raghu, S., R. C. Anderson, C. C. Daehler, A. S. Davis, R. N. Wiedenmann, D. Simberloff, and R. N. Mack. 2006. Adding Biofuels to the Invasive Species Fire? *Science* 313:1742.

Rai, V., N. Chung, M. C. Thurber, and D. G. Victor. 2008. *PESD Carbon Storage Project Database*. Stanford, Calif.: Program on Energy and Sustainable Development.

Renault. 2009. The Electric Vehicle, A Global Strategy. http://www.renault.com/en/capeco2/vehicule-electrique/pages/vehicule-electrique.aspx (accessed January 6, 2010).

Renewable Fuels Association (RFA). 2008. Ethanol Industry Statistics. http://www.ethanolrfa.org/industry/statistics (accessed January 6, 2010).

Reyes, J. F., and M. A. Sepúlveda. 2006. P<-10 Emissions and Power of a Diesel Engine Fueled with Crude and Refined Biodiesel from Salmon Oil. *Fuel* 2006:1–6.

Richey, C. B., J. B. Liljedahl, and V. L. Lechtenberg. 1980. Corn Stover Harvest for Energy Production. *Transactions of the ASAE* 25 (4): 834–39, 844.

Rípoli, T. C. C., W. F. Molina, and M. L. C. Rípoli. 2000. Energy Potential of Sugar Cane Biomass in Brazil. *Scientia Agricola* 57:677–81.

Rochelle, G. T. 2009. Amine Scrubbing for CO_2 Capture. *Science* 325:1652–54.

Rockwell, T. 1992. *The Rickover Effect: How One Man Made a Difference*. Annapolis, Md.: Naval Institution Press.

Rocky Mountain Institute (RMI). 2007. What Is a Hypercar Vehicle? http://www.rmi/org/sitepages/pid191.php (accessed March 2009).

Rose, D. J. 1974. Nuclear Eclectic Power. *Science* 184:351–59.

Rossin, A. D., and T. A. Rieck. 1978. Economics of Nuclear Power. *Science* 201:582–89.

Roush, W. 2006. Marvin Minsky on Common Sense and Computers that Emote. *Technology Review*. July. http://www.technologyreview.com/Infotech/17164/page3 (accessed January 6, 2010).

Rudin, A. 2004. How Greater Energy Efficiency Increases Resource Use. Paper presented at the North Central Sociological Association meeting, Cleveland, Ohio, April 2.

Runge, C. F., and B. Senauer. 2007. How Biofuels Could Starve the Poor. *Foreign Affairs* 86 (3): 41–53.

Sanderson, K. 2009. Wonder Weed Plans Fail to Flourish. *Nature* 461:328–29.

Schewe, P. F. 2007. *The Grid: A Journey through the Heart of Our Electrified World*. Washington, D.C.: Joseph Henry Press.

Schlesinger, W. H. 2000. Carbon Sequestration in Soils: Some Caution amidst Optimism. *Agriculture, Ecosystems and Environment* 82:121–27.

Schouten, H. 2005. Earthrace Biofuel Promoter to Power Boat Using Human Fat. *CalorieLab Calorie Counter News*. November 11. http://calorielab.com/news/2005/11/11/ (accessed December 2, 2009).

Schumacher, E. F. 1973. *Small Is Beautiful: Economics as if People Mattered*. New York: Harper and Row.

Seaborg, G. T. 1968. Some Long-Range Implications of Nuclear Energy. *Futurist* 2 (1): 12–13.

———. 1971. The Environment: A Global Problem, An International Challenge. In *Environmental Aspects of Nuclear Power Stations*, 3–7. Vienna: International Atomic Energy Agency.

———. 1972. Opening Address. In *Peaceful Uses of Atomic Energy: Proceedings of the Fourth International Conference on the Peaceful Uses of Atomic Energy*, 29–35. New York: United Nations.

———, and W. R. Corliss. 1971. *Man and Atom: Building a New World through Nuclear Technology*. New York: E. P. Dutton.

Seife, C. 2008. *Sun in Bottle: The Strange History of Fusion and the Science of Wishful Thinking*. New York: Viking.

Semenov, B. A., and N. Oi. 1993. Nuclear Fuel Cycles: Adjusting to New Realities. *IAEA Bulletin* 35 (3): 2–7.

Service, R. F. 2007. Biofuel Researchers Prepare to Reap a New Harvest. *Science* 315:1488–91.

Shinners, K. J., B. N. Binversie, R. E. Muck, and P. J. Weimer. 2007. Comparison of Wet and Dry Corn Stover Harvest and Storage. *Biomass and Bioenergy* 31:211–21.

Shnayerson, M. 2007. Quiet Thunder. *Vanity Fair*, May. http://www.vanityfair.com/politics/features/2007/05/tesla200705 (accessed January 5, 2010).

Simmons, M. R. 2009. Recent speeches and papers presented by Matthew R. Simmons. http://www.simmonsco-intl.com/research.aspx?Type=msspeeches (accessed January 6, 2010).

Sivak, M., and O. Tsimhoni. 2009. Fuel Efficiency of Vehicles on US Roads: 1923–2006. *Energy Policy* 37:3168–70.

Sky WindPower. 2009. High Altitude Wind Power. http://www.skywindpower.com/ww/index.htm (accessed January 6, 2010).

Smil, V. 1988. *Energy in China's Modernization*. Armonk, N.Y.: M. E. Sharpe.

———. 1994. *Energy in World History*. Boulder, Colo.: Westview Press.

———. 1999. China's Great Famine: 40 Years Later. *British Medical Journal* 7225:1619–21.

———. 2000. *Cycles of Life: Civilization and the Biosphere*. New York: Scientific American Library.

———. 2003. *Energy at the Crossroads: Global Perspectives and Uncertainties*. Cambridge, Mass.: MIT Press.

———. 2004. *China's Past, China's Future: Energy, Food, Environment*. London: RoutledgeCurzon.

———. 2005. *Creating the 20th Century: Technical Innovations of 1867–1914 and Their Lasting Impact*. New York: Oxford University Press.

———. 2006. *Transforming the 20th Century: Technical Innovations and Their Consequences*. New York: Oxford University Press.

———. 2008a. *Energy in Nature and Society: General Energetics of Complex Systems*. Cambridge, Mass.: MIT Press.

————. 2008b. Moore's Curse and the Great Energy Delusion. *The American: The Journal of the American Enterprise Institute.* November 19. http://www.american. com/archive/2008/november-december-magazine/moore2019s-curse-and-the-great-energy-delusion (accessed December 2, 2009).

————. 2010. *Energy Transitions.* Santa Barbara, Calif.: Praeger.

Smith, H. O., R. Friedman, and J. C. Venter. 2003. Biological Solutions to Renewable Energy. *Bridge* 33 (2): 36–40.

Smith, J., P. Smith, M. Wattenbach, S. Zaehle, R. Hiederer, R. J. A. Jones, L. Montanarella, M. D. A. Rounsevell, I. Reginster, and F. Ewert. 2005. Projected Changes in Mineral Soil Carbon of European Croplands and Grasslands, 1990–2080. *Global Change Biology* 11:2141–52

Solarbuzz. 2009a. Solar Module Price Highlights. http://www.solarbuzz.com/ Moduleprices.htm (accessed January 6, 2010).

————. 2009b. Solar Electricity Prices. http://www.solarbuzz.com/solarprices.htm (accessed January 6, 2010).

Solomon, B. D., J. R. Barnes, and K. E. Halvorsen. 2007. Grain and Cellulosic Ethanol: History, Economics, and Energy Policy. *Biomass and Bioenergy* 31:416–25.

Sørensen, B. 1980. *An American Energy Future.* Golden, Colo.: Solar Energy Research Institute.

Sparovek, G., and E. Schnug. 2001. Temporal Erosion-Induced Soil Degradation and Yield Loss. *Soil Science Society of America Journal* 65:1479–86.

Steenblik, R. 2007. *Biofuels—At What Cost? Government Support for Ethanol and Biodiesel in Selected OECD Countries.* Geneva: International Institute for Sustainable Development.

Stephanopoulos, G. 2007. Challenges in Engineering Microbes for Biofuels Production. *Science* 315:801–4.

Stephens, B. B., K. R. Gurney, P. P. Tans, C. Sweeney, W. Peters, L. Bruhwiler, P. Ciais, et al. 2007. Weak Northern and Strong Tropical Land Carbon Uptake from Vertical Profiles of Atmospheric CO_2. *Science* 316:1732–35.

Stobaugh, R., and D. Yergin, eds. 1979. *Energy Future: Report on the Energy Project at the Harvard Business School.* Cambridge, Mass: Harvard Business School.

Strauss, L. L. 1954. Speech to the National Association of Science Writers, New York City, September 16. Cited in *New York Times*, September 17, 5.

Strong, A., S. Chisholm, C. Miller, and J. Cullen. 2009. Ocean Fertilization: Time to Move On. *Nature* 461:347–48.

Sustainable Mobility Project (SMP). 2004. *Mobility 2030: Meeting the Challenges of Sustainability.* Geneva: World Business Center for Sustainable Development.

Tacon, A. G. J. 2004. *State of Information on Salmon Aquaculture Feed and the Environment.* Honolulu, Hawaii: University of Hawaii. http://www.westcoastaquatic. ca/Aquaculture_feed_environment.pdf (accessed January 6, 2010).

Teller, E., M. Ishikawa, and L. Wood. 1996. *Completely Automated Nuclear Reactors for Long-Term Operation II: Toward a Concept-Level Point-Design of a High-Temperature,*

Gas-Cooled Central Power Station System. Livermore, Calif.: Lawrence Livermore National Laboratory.

Tesla Motors. n.d. Reducing Dependence on Foreign Oil. http://www.teslamotors.com/learn_more/foreign_oil.php (accessed January 5, 2010).

Thomas, S. R. 2003. *Causes and Effects of Variation in Corn Stover Composition.* Golden, Colo.: National Renewable Energy Laboratory.

Tillerson, R. W. 2008 Meeting Global Energy Supply and Demand Challenges. Remarks at the 19th World Petroleum Congress, Madrid. July 1. http://www.exxonmobil.com/Corporate/news_speeches_20080701_RWT.aspx (accessed January 6, 2010).

Tollefson, J. 2009. Charging the Future. *Nature* 456:436–40.

Tréguer, P., and P. Pondaven. 2000. Silica Control of Carbon Dioxide. *Nature* 406:358–59.

Trumbore, S. E., and C. I. Czimczik. 2008. An Uncertain Future for Soil Carbon. *Science* 321:1455–56.

United Nations. 2008. *World Population Prospects: The 2008 Revision.* http://esa.un.org/unpp (accessed January 6, 2010).

United Nations Food and Agriculture Organisation (FAO). 2003. *State of the World's Forests 2003.* Rome: United Nations Food and Agriculture Organisation.

U.S. Atomic Energy Commission. Division of Technical Information. 1971. *Breeder Reactors.* By W. Mitchell and S. E. Turner. *Understanding the Atom Series.* Washington, D.C.: Government Printing Office.

U.S. Bureau of Economic Analysis. 2008. U.S. GDP, 1930–2008. http://www.bea.gov/national/xls/gdpchg.xls (accessed January 6, 2010).

U.S. Census Bureau. 2009. U.S. Population Projections. http://www.census.gov/population/www/projections/natproj.html (accessed January 14, 2010).

U.S. Congressional Budget Office. 2009. *The Impact of Ethanol Use on Food Prices and Greenhouse Gas Emission.* April. http://www.cbo.gov/ftpdocs/100xx/doc10057/04-08-Ethanol.pdf (accessed December 16, 2009).

U.S. Congressional Research Service. Resources, Science, and Industry Division. 2003. *Automobile and Light Truck Fuel Economy: The CAFE Standards.* By R. Bamberger. Order code IB90122. March 12. http://www.ncseonline.org/NLE/CRSreports/03Apr/IB90122.pdf (accessed January 6, 2010).

U.S. Department of Agriculture. 2004. *The 2001 Net Energy Balance of Corn Ethanol.* By H. Shapouri, J. A. Duffield, and M. Wang. http://www.usda.gov/oce/reports/energy/net_energy_balance.pdf (accessed January 6, 2010).

———. 2007. Nitrogen Used on Corn. http://www.ers.usda.gov/Data/FertilizerUse/Tables/Table10.xls (accessed January 6, 2010).

———. Forest Service. 2005. *A Summary and Comparison of Bird Mortality from Anthropogenic Causes with Emphasis on Collisions.* By W. P. Erickson, G. D. Johnson, and D. P. Young. USDA Forest Service General Technical Report PSW-GTR-191. 1029–42. http://www.fs.fed.us/psw/publications/documents/psw_gtr191/Asilomar/pdfs/1029-1042.pdf (accessed December 16, 2009).

U.S. Department of Energy. 2007. DOE Selects Six Cellulosic Ethanol Plants for Up to $385 Million in Federal Funding. Press release. February 28. http://www. energy.gov/news/4827.htm (accessed January 6, 2010).

————. 2008. *Carbon Sequestration Atlas of the United States and Canada*. Morgantown, W. Va: National Energy Technology Laboratory. http://www.netl.doe.gov/ technologies/carbon_seq/refshelf/atlas (accessed January 6, 2010).

U.S. Department of Energy. Energy Information Administration. 2009. *Annual Energy Review 2008*. http://www.eia.doe.gov/emeu/aer/overview.html (accessed January 6, 2010).

U.S. Department of Transportation. Bureau of Transportation Statistics. 2007. *National Transportation Statistics*. http://www.bts.gov/publications/national_ transportation_statistics (accessed January 14, 2010).

U.S. Department of Transportation. Research and Innovative Technology Administration. Bureau of Transportation Statistics. 2007. *National Transportation Statistics*. http://www.bts.gov/publications/national_transportation_statistics (accessed January 6, 2010).

U.S. Geological Survey. 2000. *World Petroleum Assessment 2000*. http://pubs.usgs. gov/dds/dds-060 (accessed January 6, 2010).

Van der Zwaan, B. C. C. 2002. Nuclear Energy: Tenfold Expansion or Phase-Out? *Technological Forecasting and Social Change* 69:287–307.

Vance, E. 2009. High Hopes. *Nature* 460:564–66.

Vestas. 2007. 3.0 MW—An Efficient Way to More Power. http://www.vestas.com/ en/wind-power-solutions/wind-turbines/3.0-mw.aspx (accessed January 14, 2010).

Voelcker, J. 2007. Top 10 Tech Cars 2007. *IEEE Spectrum* 44:34–41.

Von Hippel, F., and S. Jones. 1997. The Slow Death of the Fast (Plutonium) Breeder (Reactor). *Bulletin of the Atomic Scientists* 53 (5): 46–51.

Von Lampe, M. 2006. *Agricultural Market Impacts of Future Growth in the Production of Biofuels*. Paris: Organisation for Economic Cooperation and Development.

Walker, J. S. 2004. *Three Mile Island: A Nuclear Crisis in Historical Perspective*. Berkeley, Calif.: University of California Press.

Walter, K. M., S. A. Zimov, J. P. Chanton, D. Verbyla, and F. S. Chapin. 2006. Methane Bubbling from Siberian Thaw Lakes as a Positive Feedback to Climate Warming. *Nature* 443:71–75.

Weinberg, A. M. 1973. Long-Range Approaches for Resolving the Energy Crisis. *Mechanical Engineering* 95 (6): 14–18.

————. 1994. *The First Nuclear Era: The Life and Times of a Technological Fixer*. New York: American Institute of Physics Press.

White House. 2009. President Obama Announces National Fuel Efficiency Policy. Press release. May 19. http://www.whitehouse.gov/the_press_office/President-Obama-Announces-National-Fuel-Efficiency-Policy (accessed January 6, 2010)

Wigley, T. M., and D. S. Schimel, eds. 2000. *The Carbon Cycle*. Cambridge: Cambridge University Press.

Wilson E. J., M. G. Morgan, J. Apt, M. Bonner, C. Bunting, J. Gode, R. S. Haszeldine, et al. 2008. Regulating the Geological Sequestration of CO_2. *Environmental Science and Technology* 42:2718–22.

Workshop on Alternative Energy Strategies (WAES). 1977. *Energy Supply–Demand Integrations to the Year 2000*. Cambridge, Mass.: MIT Press.

World Trade Organization (WTO). 2009. *International Trade Statistics 2009*. Geneva: World Trade Organization.

Index

About the Author

Vaclav Smil is a Distinguished Professor in the Faculty of Environment at the University of Manitoba in Winnipeg and a Fellow of the Royal Society of Canada. His interdisciplinary research has included the studies of energy systems (resources, conversions, and impacts), environmental change (particularly global biogeochemical cycles), and the history of technical advances and interactions among energy, environment, food, economy, and population. He is the author of thirty books and more than three hundred papers on these subjects and has lectured widely in North America, Europe, and Asia.